Smoothing, Filtering and Prediction - Estimating The Past, Present and Future

Smoothing, Filtering and Prediction - Estimating The Past, Present and Future

Editor

Jeremy Weissberg

scitus
academics

Smoothing, Filtering and Prediction - Estimating The Past, Present and Future
Edited by **Jeremy Weissberg**

ISBN: 978-1-68117-606-2
Library of Congress Control Number: 2016946461

SCITUS Academics LLC,
616, Corporate Way, Suite 2, 4766,
Valley Cottage, NY 10989

www.scitusacademics.com

Printed in United States of America on Acid Free Paper

Preface

Smoothing is often used to reduce noise within an image or to produce a less pixelated image. Most smoothing methods are based on low pass filters. Smoothing is also usually based on a single value representing the image, such as the average value of the image or the middle (median) value. In image processing, to smooth a data set is to create an approximating function that attempts to capture important patterns in the data, while leaving out noise or other fine-scale structures/rapid phenomena. In smoothing, the data points of a signal are modified so individual points (presumably because of noise) are reduced, and points that are lower than the adjacent points are increased leading to a smoother signal. Smoothing may be used in two important ways that can aid in data analysis; by being able to extract more information from the data as long as the assumption of smoothing is reasonable and; by being able to provide analyses that are both flexible and robust. Filtering and prediction is about observing moving objects when the observations are corrupted by random errors.

Smoothing, Filtering and Prediction - Estimating The Past, Present and Future describes the classical smoothing, filtering and prediction techniques together with some more recently developed embellishments for improving performance within applications. It aims to present the subject in an accessible way, so that it can serve as a practical guide for undergraduates and newcomers to the field.

Table of Contents

CHAPTER 1

Approach Development Accelerate of Process Special Traffic Filtering

Karimov Madjit Malikovich, Gulomov Sherzod Rajaboevich*, Yusupov Bakhodir Karomatovich

Department of Information Security, Faculty of Computer Engineering, Tashkent University of Information Technologies, Tashkent, Uzbekistan

ABSTRACT

This article is dedicated to the analysis list of a set of rules to traffic filtering, which is a multi-di- mensional structure, where each dimension is a set of networking field or the field of action, measuring the cost of the rules to traffic filtering on computer networks, allowing to determine the difference between definition of the rules and the control of the packet fields. Furthermore, the article was considered a hierarchical model to optimize traffic filtering, which reduces the overhead traffic filtering rules and provides the semantic integrity of the original set of rules to traffic filtering. The hierarchical structure of the design and optimization of traffic filtering was researched. And also was developed the hierarchical approach to optimize traffic filtering for reducing set of rules traffic filtering. Analyzed the algorithm optimal solutions and algorithm of random search filters that, allowing you to find the shortest way to a set of rules to traffic filtering. Moreover, in this article was presented the effectiveness evaluation of the process accelerating traffic filtering proposed by HAOTF.

KEYWORDS

Traffic Filtering, Hierarchical Structure, Model, HAOTF, Heuristic Method

1. INTRODUCTION

Integrated efficiency, reliability and availability of filtering software are crucial in ensuring the security and administration, especially when the network is under attack and threat. These problems require new designs, architecture and algorithms to accelerate traffic filtering. With the dynamic change of the load on the network topology and bandwidth demand, traffic filtering becomes a vulnerable point. All these factors create a demand for the most efficient, high-performance, affordable and reliable traffic filtering in Firewalls. Nowadays, Firewall enforces security policy with a set of multi-dimensional traffic filtering rules. Information security policy in networks is a very important task to speed up traffic filtering. In addition, with the increase in capacity of existing networks for processing and forwarding traffic at extremely high speed, Firewalls are very limited resources. Thus, the main objective is to overcome the drawbacks of the current traffic filtering and enhance their ability to dynamic changes in the load and network topology, especially in attack. The main objective of these rules is to maintain the semantic integrity policy established at each level of hierarchical models.

2. LIST A SET OF TRAFFIC FILTERING RULES

Security rule is a multidimensional structure, where each dimension is a set of network fields or field of action. A set of rules defines a security policy that traffic filtering. Rules define by a set of IP-source address and a set of destination IP-address a variety of service types and fields of action. The type of service usually includes a main type of protocol and port number. Field of action can either accept or refuse or direct. Take action allows access packet in a secure domain. Disclaimer action causes the packet, in violation of security policy. Formal rules R can be represented as: $R=\left[\Phi^1,\Phi^2,\cdots,\Phi^k;\Sigma\right]$, where Φ is the network field and the field of action. An example of a typical rule may be the following sequence:

$$< src=\{s_1,s_2,\cdots,s_n\};dst=\{d_1,d_2,\cdots,d_m\};srv=\{\varsigma_1,\varsigma_2,\cdots,\varsigma_l\}$$
$$\text{action}=\{\text{drop / accept / forward}\}> \quad (1)$$

where s_n represents IP-address of the source, d_m destination IP-address and ς_l type of service. In the list of a set of rules to filter traffic, rules describing network security policies, formed as a "priority list". Priority rules, also referred to as the rules of rank, based on its position within the list [1] . Previously derived rules are ranked higher than those which were later. The set of rules may be transformed by creating all possible permutations of the fields in a rule. Formal representation of a set:

$$< src=s_n;dst=d_m;srv=\varsigma_l$$
$$\text{action}=\{\text{drop / accept / forward}\}> \quad (2)$$

List a set of rules to traffic filtering work by studying sets in sequential order. For each packet, the first matching set determines the action of traffic filtering. If the size of the list is small, then the list of a set of rules to traffic filtering is considered good. Since the size of the list are scaled up to one million sets, management and optimization of traffic filtering policy is a challenge.

2.1. Measuring the Cost of Traffic Filtering Rules

The main factor, which influences the performance of the filtering action of traffic are overhead data for packet inspection. Measurement computation is performed according to the rule, and can be readily applied to a set of rules within. To determine the overheads incurred by the traffic filtering treatment rules and application security policy, there are two specific measurements.

The first dimension, referred to as the size of the rules. The size of the rule in terms of the number of bits necessary to determine unambiguously as the difference between the definition of rules and control of the field of the packet. Assumptions underlying the measurement of the size of the rules related to the fact that the complexity of coordination of operations proportional to the size rules. Formally, this rule r, rules size (r) can be defined as:

$$\text{Rulesize}(r) = \left\{\sum S_p, D_p \left\{\alpha_1 \times \|s_p\| + \alpha_2 \times \|d_p\|\right\} + \beta \times N_s \times \left(\|Pr_r\| + \|Pr_o\|\right)\right\} \tag{3}$$

where α_1, α_2, и β are main parameters, S_p, и D_p respectively, a set of source and destination prefixes that occur in the rule definition, s_p, и d_p respectively are bit representation of the source and destination prefixes, N_s the number of services defined in rule, и Pr_r, и Po_r, are bit representation of the protocol and ports.

The second dimension used in the experiment of current costs at a given set of rules. These costs depend on the size and positioning of the rules on how often the rule is called traffic filtering. Formally, given a set of rules $r_1, r_2, \dots, r_k$, costs of this rule r_i, is defined as follows:

$$\text{Expenditure}(r_i) = \text{number of retries}(r_i) \times \sum \forall r_k \in Pr_i \|r_k\|$$

where Pr_i a set r_i-espredecessors on the basis of a list of a set of rules. Using the above measurements in order to optimize the reduction in size and therefore the rule set, the processing time rule set. This, in turn, reduces the overall operational costs of traffic filtering.

2.2. Criterion Conversion of Traffic Filtering

Traffic filtering defines its security policy with a set of policies or security rules. These security policies are defined formally as follows:

Let F is a list of the original set of rules to filter traffic.

Let $T(F)$ is converted traffic filtering, which retains the properties and a set of rules F. Is determined by the cost function, $C(f)$, which is the average overhead costs set of rules to filter traffic.

$T(F)$ is acceptable conversion F:

$T(F)$ retains the properties and rules F (semantic integrity of the property);

$C(T(F)) \leq C(F)$ (property costs).

2.3. Transformation Approach of Traffic Filtering

Converting traffic filtering is achieved in the process of dividing the original set list of rules to filter traffic in the rule group subsets. A set of rules for filtering traffic sorted based on the traffic characteristics before conversion initiated. Getting the right subset retains the properties and rules of "semantic integrity" of the original list, a set of rules. In order to formulate a formal:

Let $S(F)$ is an approach conversion filter traffic based on a set of rule F.

$S(F)$is acceptable conversion F, if:

$S(F)$retains the properties and rules F (semantic integrity properties);

costs (S(F)) £ performance F (property costs).

Semantic integrity of the property:

Let there be at least one rule r in $S(F)$ such that the action r network p packet is different than F on p. This means that the rule of traffic filtering installed F on this network traffic is different from the actions taken by the new transformed representation $S(F)$.

Thus, the semantics of the original set of rules to traffic filtering is not equivalent to transformation a set of rules to traffic filtering.

Rules based on a list of a set of rules to traffic filtering is scanned in a sequential order and divided into subsets of rules based on traffic characteristics. The order, in which the rules are divided into subsets of rules, does not add a new rule, or cause the removal of any rules. This means that no new rules are created or removed from the conversion process. This proves that the rule r in $S(F)$ must belong to some level of priority in the set of rule F, which implies a contradiction.

Consequently, there are no rules to convert a set of rules to traffic filtering $S(F)$, which have different activities, unlike those that exist in the original set of rules F. This proves that the semantic properties and rules in the original set of rules stored after conversion.

Property costs:

Let the overhead costs of the original set of rules to filter traffic and a set of transformation rules will be $C(F)$ and $C(S(F))$, respectively.

It is suggested that $C(F)>C(S(F))$.

For this assumption to be true, there exists at least one packet p, that it complies with the rule r a set of rules to traffic filtering, where compliance costs in the transformed set of rules to traffic filtering higher than the corresponding cost in the original set of rules.

Let overheads processing network packet P, that matches the rule, r a set of rules F traffic filtering is represented as x and the relevant rules r transformed traffic filtering $S(F)$ there will be y, where $y>x$.

Thanks to the list of rules to traffic filtering capacity y would be more than x, if the series r in $S(F)$ lower than the series of r in F.

Since both sets of rules to traffic filtering characteristics are sorted traffic (inbound), there is a rule r' in the rule subset $S(F)$ traffic filtering, which has a higher number than the r of the low number of the original set of rules F.

So, all the rules in F sorted in accordance with the calculation information and no new rules created or deleted as a result of the conversion process.

Thus, it have proved that the costs set of rules $S(F)$ traffic filtering is £ on the cost of the original set of rule F. In the worst case, the length $S(F)$ will be equal to the length F, meaning $C(F)=S(F)$.

3. THE HIERARCHICAL STRUCTURE OF THE OPTIMIZATION OF TRAFFIC FILTERING

Data structure. Multilevel data structure consists of subsets of rules and their respective filters. To process typically uses a hierarchical data structure, in which a deep level of the hierarchy contains a subset of the rule and intermediate levels contain filters that cover the rules that are included in these subsets. Balances the hierarchical structure in order to reduce the length of the deepest subset rules, therefore, it is important to have the desire to achieve the maximum reduction of the costs of processing rules [2] . The data structure must ensure a reduction of overhead costs, semantic integrity and safety of the initial set of rules. It should be noted that overhead costs are determined by the rule subset. Furthermore, the data structure must be designed so that the process of re-balancing, in response to changes in traffic can be achieved at minimal cost.

Semantic integrity of the original set of rules can be achieved, and set during the reign of the separation pro- cess by calculating the filter that is clearly and fully complies with the rule subsets.

Moreover, after the packet processing must follow the same semantic instructions filters produced in the separation process. If the rules are divided and re-loaded, then to optimize the overhead costs, the process of repeating the original semantics of the rules must be achieved with lower costs.

3.1. The Hierarchical Development Model

The hierarchical model defines approach to designing networks and includes three logical layers (seeFigure 1):

- corelayer;
- distribution layer;

- access layer.

Each layer has its own function. Three layers do not necessarily imply the presence of three different devices. To draw an analogy with the hierarchical model of OSI, it is a separate protocol does not always correspond to one of seven levels. Sometimes protocol corresponds to more than one level of OSI model, and sometimes several protocols implemented within the same layer. And in constructing hierarchical networks, at one layer can be multiple devices or a single device that performs all functions defined on two adjacent layers.

The core layer is at the top of the hierarchy and is responsible for the safe and rapid transfer of large amounts of data. Traffic transmitted through the core, is common to most users. Most user data is handled at the distribution layer, which, if necessary, forwards the request to the core. For core layer its importance has resiliency as a failure at this layer could result in loss of connectivity between the levels of the distribution network.

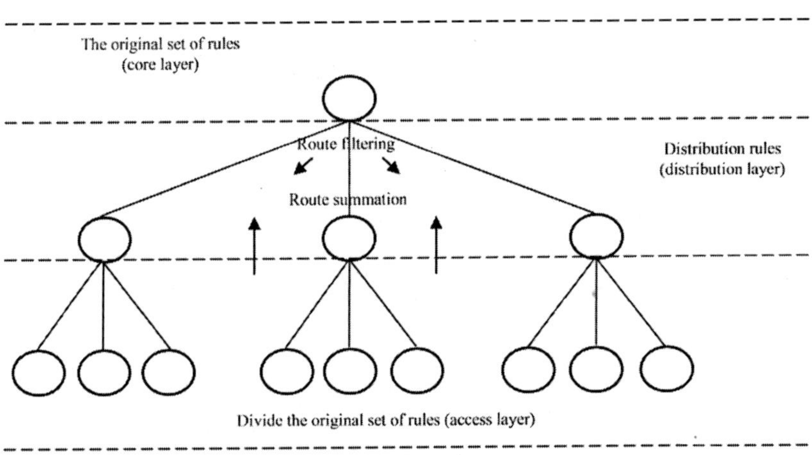

Figure 1. The layers of hierarchical structure.

The core layer for the following functions:

- shared bandwidth;
- switching bandwidth;
- MAC-level filtering;
- microsegmentation.

The distribution layer, which is sometimes called the working group layer, is the link between the layers of access and core. Depending on the implementation method, the distribution layer can perform the following functions:

- verification of source and destination addresses;
- check the input and output ports;
- filter routes to hide the internal network;
- providing routing, quality of service and network security;
- canal aggregation;
- transition from one technology to another (e.g. 100Base-TX to 1000Base-T);
- aggregated bandwidth low-bandwidth access to the high-speed backbone links.

The access layer controls user access and working groups to resources united network. The main objective access layer is to create points of exit/enter users to the network. This layer performs the following functions:

- continued (from the level of the distribution), access control and network policy;

- create separate collision domains (segmentation);
- connecting workgroups to the level of distribution;
- access layer uses switched LAN.

And so, the core layer is required in order to produce a separation of the original set of rules to a set of disjoint subsets of rules. This process involves taking a set of access rules and rules on the division of its subsets, each of which is determined by the filter. Each filter consists of a series of disjoint sets that fully cover its corresponding rule subset (see Figure 1).

At distribution layer, rule is distributed in such a way that high standards of cost have been moved to the top of a set of rules. As previously stated, the rules are based on cost rules size. The amount of traffic passing through this rule indicates the number of his matches. At this stage, costs are reduced by the general rules of traffic processing.

The access layer is taken as the input of the original list a set of rules based on the list, and produces an optimized set of rules. This is the best set of rules consists of fully overlapping and concise rules, where all exceptions to the rules and dependencies are removed. The fact that the rules in the rule set intersect to change the order of the rules and dividing them into subsets of rules without breaking the semantics of the original set of rules at the kernel level.

3.2. Development of a Hierarchical Approach to Optimize Traffic Filtering

To develop a hierarchical approach to optimize traffic filtering (HAOTF) uses a hierarchical structure to the division of the initial set of rules and regulations mutually exclusive subsets to reduce the overhead costs of traffic filtering.

In the original set of rules, HAOTF uses a multistep process in which it originally shared the original rule set into two subsets. Then recursively passes the access layer to subsets produced in the previous step generate the next layer of the hierarchy. Formation to access layer continues as long as the total cost of processing does not eclipse the benefit gained by continuing the current access layer subsets.

Then, in this case, the access layer is terminated and the previous layer is selected as optimum as possible the depth of hierarchical structure. Efficiency distribution layer depends strongly of a subset of the rules used at different layers of the hierarchy.

In solution the complex markup, HAOTF uses an iterative approach for the separation of the original set of rules and make a multilevel hierarchy, economically balanced subsets of rules.

Initially, the set of rules is divided into two subsets and the filters are designed and cover rules contained in each subset.

In HAOTF the time of packet processing begins at the root of the hierarchical structure. Packets subsequently are sent to the other layers of the hierarchy for further processing. Packet processing is terminated, if the difference between the attributes of the packet, as defined in the traffic filtering, the security policy is triggered. In this case, the action is determined by compliance with the relevant filter rules. The default action may be taken in either case the packet is sent to the destination, or rejected, and in this case, the packet is dropped. Given a large set of rules, the goal HAOTF is to partition the set of mutually exclusive subsets K. Each subset is associated with a unique filter which is an extension of associated policies subsets. HAOTF driven by three major design goals:

- to reduce the cost of processing a set of rules determined by the average processing time packet, which he takes on;
- to save the semantics of the original set of rules;

- to maintain an optimal set of rules as templates for traffic and changes the set of rules.

It should be noted that in the general form of K may be reduced to the “clustering”. Figure 2describes action rules N in subset K.

The resulting subsets, along with according filters, form a first layer of the hierarchy. This iterative process continues until, while further division of the subsets at the current level of the hierarchy will not be effective. It should be noted that this cost also includes certain expenses filters. The structure of the basic steps described on HAOTF in Figure 3.

3.3. Service Hierarchical Structure

The hierarchical structure is built based on the current traffic patterns and a set of rules. As traffic patterns and a set of rules changes, the hierarchical structure should be updated to maintain its balance.

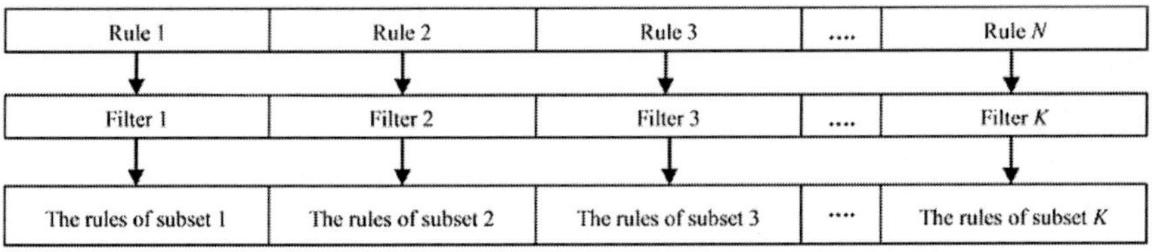

$$\text{Filter}_i \cap \text{Filter}_j = \phi, N \gg K$$

Figure 2. Action rules N in subset K.

To identify changes, HAOTF monitors traffic logs in real time and adjusts counters. HAOTF notifies that changes have occurred, if the difference between the old and the updated graphs of any rules exceeds a predetermined threshold.

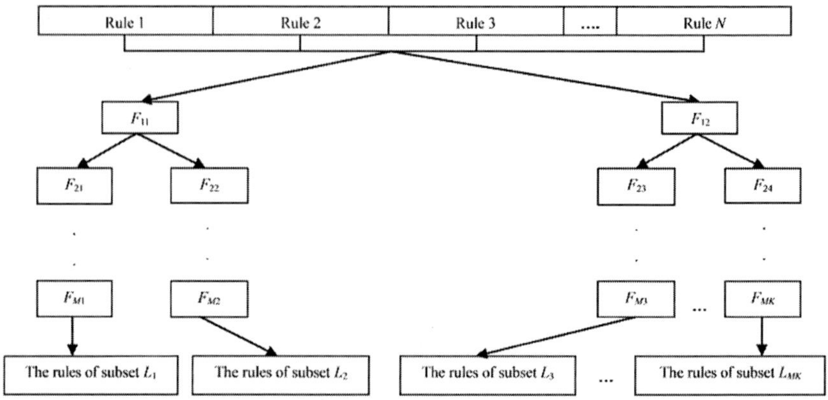

$$F_i \cap F_j = \phi, N \gg K$$

Figure 3. The basic action structure of HAOTF.

This threshold is configurable parameter that is determined based on the traffic characteristics. If necessary to balance the hierarchical structure, HAOTF uses existing cost upgrade path rules in the rule of subsets, including the rules that have been added to reflect the new security policy. Then HAOTF uses reverse distribution layer, reverse access layer and also assistance layer in restoring balance hierarchical structure. The flowchart of HAOTF between layer hierarchical structures is shown in Figure 4.

The reverse distribution layer consists of redistributing priorities in the rule subset deeper layer hierarchical structure. This layer takes effect variations in traffic to the number of rules in the rule set. The reverse distribution layer is triggered when between current and previous number of rules exceeded the threshold value. The reverse access layer caused, when the hierarchical structure becomes out of balance due to changes in traffic [3] . The hierarchical structure is considered to be out of balance, if the average cost of processing packets exceeds a predetermined threshold. This layer may be at any layer, including at the root of the hierarchical structure. When the balance is a hierarchical structure, the access layer is applied to the initial subset rule that is generated from the hierarchical structure. In some cases, impossible to obtain a more balanced hierarchical structure, in this case indicated as optimal layer increases and the threshold for the intermediate layers. The assistance layer aimed at reducing overhead costs packet processing at different layers of the hierarchy. The need to promote the rules occurs when one number rules sharply increases and exceeds a predetermined threshold value. This scenario may occur during abnormal traffic behavior, often observed in a denial of service and distributed denial of service (DoS/DDoS) attacks. To mitigate the impact of DoS/DDoS attacks to significantly reduce the cost of processing the traffic generated by these attacks, usually rises to a layer above the filter [4] . Depending on the rules, the assistance layer may continue recursively until it reaches an appropriate layer of priority. In extreme cases, the rules can be moved all the way up to the root of the hierarchical structure.

This assistance is temporary and usually it is never removed from the rules of subsets. The reason is due to the temporary facilitate temporary state of DoS/DDoS attacks. After the traffic returns to its normal layers, may typically be removed from the higher layers.

The “match” function—checks that the set is covered by the filter. Source and destination IP-addresses are compared with the range specified in the filter. The same port number is compared with a range of ports in the filter. Protocol type is checked against a list of protocol types. This function returns if they meet the set, otherwise they are false.

The “distance” function calculates the distance, given a set of filter. If the filter corresponds to the set, the value returned by this function 0. Otherwise, this function returns 1, positive number between 0 and 1, not inclusive. If the IP-address ranges are then calculated based on a distance function between the two most distant points within. A similar procedure is used to calculate the distance between ports or port ranges. Distance protocol 0, if the protocol already exists in the list of protocols for the filter. Otherwise, sets the distance 1. All distances are normalized to the maximum value corresponds to their fields. Summation of these normalized values are compared again and normalized to produce a value between 0 and 1.

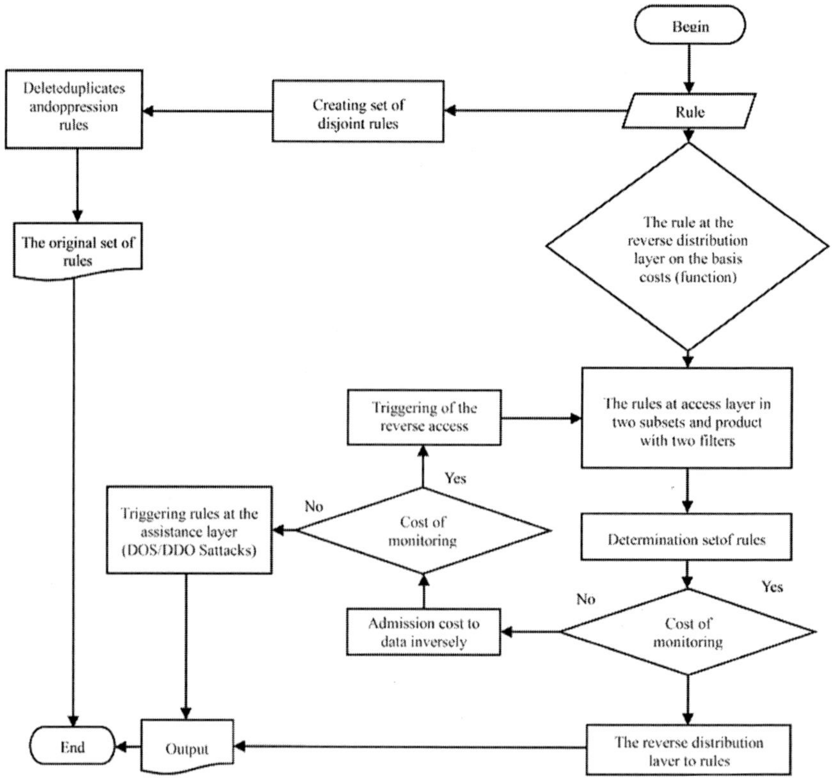

Figure 4. Flowchart of HAOTF between layer hierarchical structures.

The "extension" function is used to extend the filter so that it corresponds to a given set. This is achieved by expanding the range of

IP, port range and protocols. The function calculates costs based on a set of traffic characteristics and other property sets (see Figure 5).

4. THE ALGORITHM OF OPTIMAL SEARCH SOLUTIONS

The algorithm of optimal search solutions uses a priori estimates of the cost of the path to the target state, which ensures high efficiency of the search. The algorithm assumes that there are two lists of vertices of open and closed. In the first peaks are not yet proven algorithm and the second those vertices that have already met in the search for solutions. At each new step from the list of open vertices selected vertex with the smallest weight.

The basic idea of the algorithm is used for each node n on the graph of the state space of the evaluation function of the form $f(n)=g(n)+h(n)$. There $g(n)$ on the graph represents the distance from the node n to the initial state, and $h(n)$ estimate of the distance from the node n to the node representing the final (target) condition. The lower value of the evaluation function $f(n)$ better, i.e. node n lies at a short distance from the initial state to the target. The idea of the algorithm is to via $f(n)$ find the shortest path on a graph from the initial state to the target (Table 1).

Achieving the algorithm of optimal search solutions in HAOTF possible, since the costs calculated in the aggregate for each division as fixed and does not vary with the priority set of rules.

There function $g(n)$ determines the cost of a set of filter rules from node n to the initial state.

Function $h(n)$ on the other hand, calculates the optimal cost remaining sets assigned if placed in any of the subsets, then the value is an approximation of the path to the target node n.

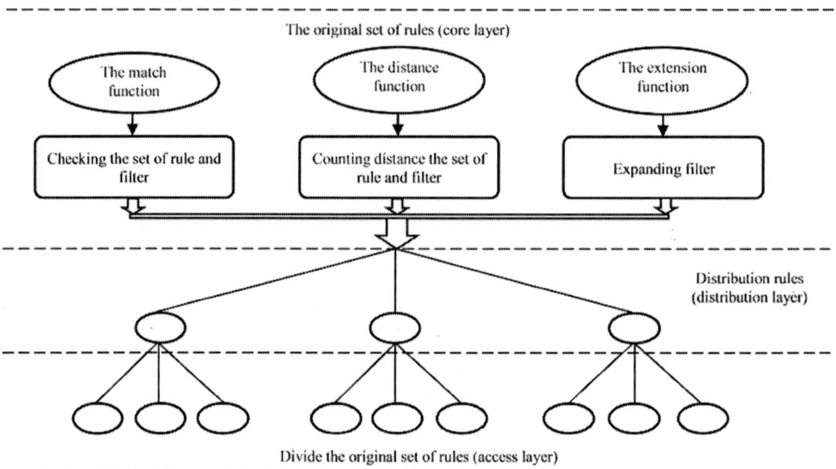

Figure 5. The distributionfunction ofthe core.

Function $h_{max}(n)$ calculates the maximum number of remaining sets.

It can be used as a guide to complete the calculation of the filter, if the value of the benefits arising from the new filters do not increase the benefits of the previous configuration. Another mechanism that is used to reduce the overhead costs incurred in the search for a possible optimal solutions, limiting the search space.

It is triggered when the difference between $h_{max}(n)$ and $h_{min}(n)$ lower than the specified error rate. It is convenient to search for filters and is an optimal solution with a much faster pace.

$$f(n)=g(n)+h_{max}(n)/h_{min}(n) \tag{4}$$

Figure 6 shows a topological model search filters in action, where the nodes—is a list of a set of rules and $h(n)$ is the shortest distance to the initial state. Red color—the initial state, blue color—the goal, orange color—visited nodes (see Figure 6).

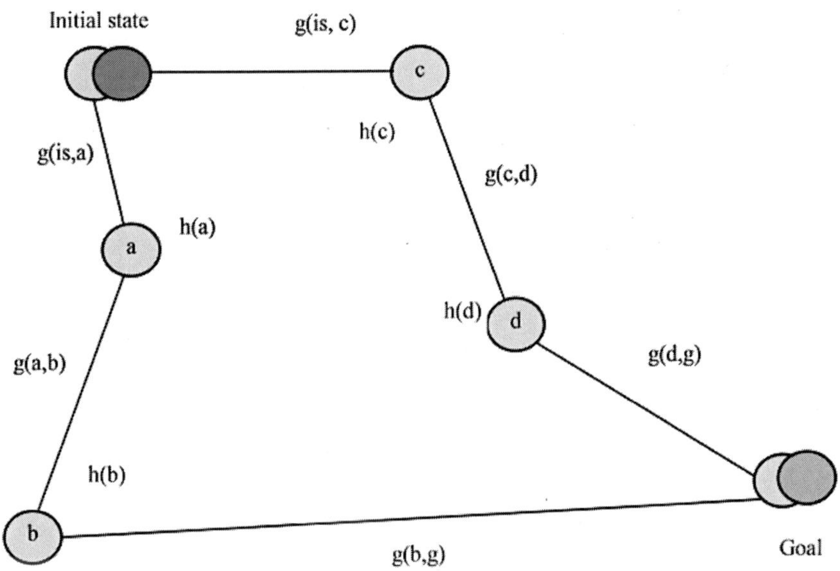

Figure 6. Topological model of searchfilters.

Table 1. Callating algorithm search of path filters.

1.	$f(a) = g(is.a) + h(a)$
2.	$f(c) = g(is.c) + h(c)$
3.	$f(b) = g(is.a.b) + h(b)$
4.	$f(d) = g(is.c.d) + h(d)$
5.	$f(g1) = g(is.a.b.g) + h(g)$
6.	$f(g2) = g(is.c.d) + h(g)$

Basic steps algorithm of search filters was shown in Listing 1.

Listing 1.

$g(i,n)$= costs $list_a$ и$list_b$ after addition of n sets of the list

$h(n)$= current cost optimal placement of the number of sets $(i,n)=g(i,n)+h(n)$

$filetr_a$, $list_a$ = filter and set for list A

$filter_b$, $list_b$ = filter and set for list B

steck = the required stack and the smallest upper value.

Input:

set[] - List sorted sets cost

Algorithm

Counter = 0; $list_a$= θ, $list_b$=θ

the current set = set[counter]

until the counter < set size ()perform

if costs (A, current set) < of expenses (B, current set)

then

iff$ilter_a$ $filter_b$expansion (the current set) ≠

then

createsteck(<$list_a$, listbcurrentset, filtera, filterb. expansion (current set),counter>)</list

End If

filteraexpansion (current set), filtera .create(currentset)

iffilteraand filterb =

then

<lista, listb, filtera, filterb, counter> = steck none()</list

End If

End If

counter

current set = set[counter]

end as long as

Out:<filtera, lista, filterb, listb></filter

Despite the improved algorithms optimal search solutions defined drawback of this algorithm (seeFigure 7). The disadvantage is that the memory requirements may also be unacceptable, as the number of sets becomes large. To solve this drawback is proposed heuristic method of random search.

5. HEURISTIC METHOD OF RANDOM SEARCH

Heuristic method of random search based on the sequence of random generation of a large number of values in between finding the roots followed by narrowing these gaps.

The main steps of the method of random search filters:

- in each series with a random number generator is formed from an array of N points of the function $F(x_i), x_i \in [x_{iM}, x_{iH}]$ $(i = 1, 2, \cdots, n)$; n-number of filters $(N > 100)$.
- among the elements of the array of the function is the optimal value $(V_{min}|V_{max})$, as well as the corresponding value of the variable $(X_{min}|X_{max})$.

And so forth:

Typically, the choice of solutions can be represented sequence elections. If you make this election with the help of a random mechanism, the solution is very fast, so that we can find a solution several times and memorize the "record", the best of solutions encountered. The proposed method of random search filter is a coarse search for solutions aimed at defining a set of filters and separation set list set into two subsets. Each set of list of bases independently of the other set.

- is calculated for each new coordinate interval, within which is produced after selection of the N filter.

And so, in order to reduce costs on a set of rules 10%, $M = 0.9 * (b - a)$

$$
\begin{aligned}
&a1 = x - M/2; \text{ if } a1 < a \text{ then } a1 = a \\
&b1 = x + M/2; \text{ if } b1 < b \text{ then } b1 = b
\end{aligned}
\tag{5}
$$

Figure 8 shows a topological model of random search filters in action, here:

The starting point of the state of the filter represented by the vector x_0, declared its current and calculated values of the objective function it.

- The current state of the filter is attached to the point increment in the form of a random vector delta and calculated value of the objective function.

- If the value of the objective function improved, then this point is the current.

- Check the stop condition. If it holds, then go to step 4, Otherwise, in a step 2.

- if $(x_0 = x_1)$ и $(x_1 = x_2)$, while improving the objective function;

- if ($x_0 = y_1$ или $x_0 = y_2$) и ($x_1 = y_3$ или $x_1 = y_4$), then the objective function is deteriorating or remains the same.

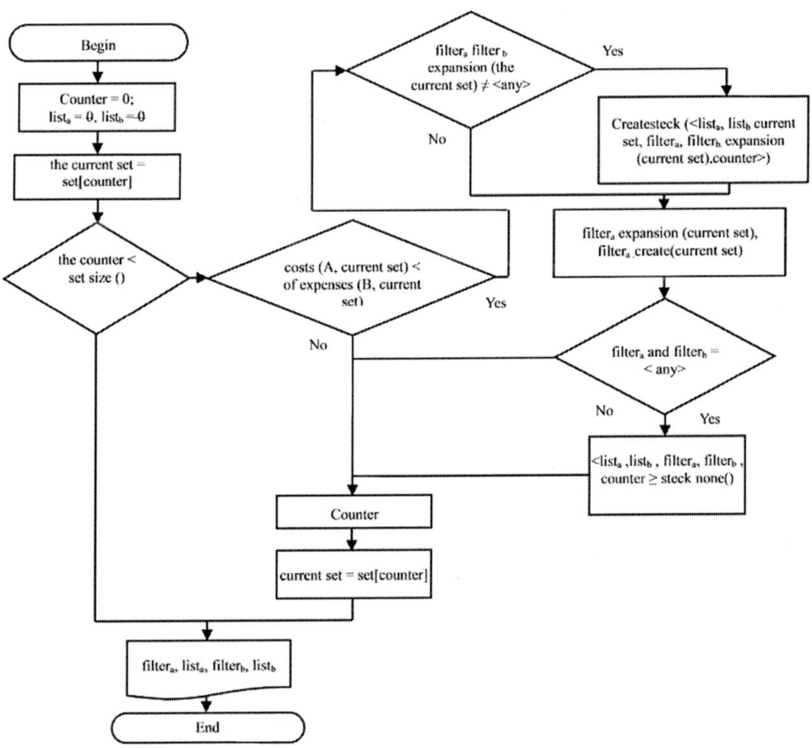

Figure 7. Flowchart of algorithmoptimal search solutions.

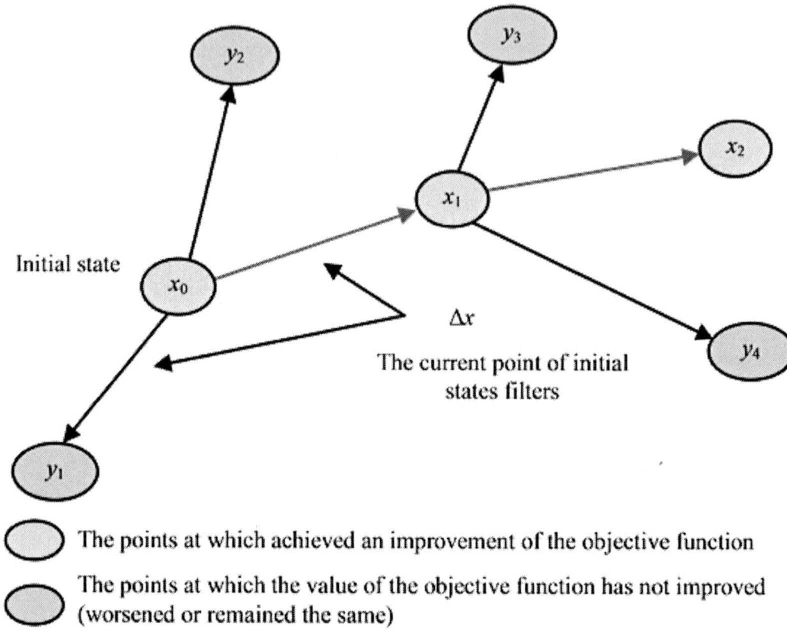

Figure 8. Topological modelof random searchfilters.

This division will perform sets into smaller sets of subsets. This method uses a randomly chosen initial condition of the filter and uses a random algorithm (see Listing 2) and search strategy for determining an initial filter from a set of possible filters (see Figure 9).

In applying the method sets do not intersect with each other. In other words, the creation of a set of non- overlapping with to each other [5], allow complete flexibility of access layers and reverse distribution based on traffic characteristics.

Basic steps algorithmof random search filters was shownin Listing 2 (see Figure 10).

Listing 2.

```
Input:

set[] - lists set sorted by expenditure

filtera = set[0]

filterb = set[1]

for i = 2 length set() do

iffilteramatches (set[i])

then

create set[i] for lista

else

iffilterbmatches (set[i])

then

create set[i] for listb

else

distancea =filteradistance(set[i])

distanceb =filterbdistance(set[i])

ifdistancea<distanceb</distance

then

filteraexpansion(set[i])

create set[i] for lista
```

else

filterbexpansion(set[i])

create set[i] for listb

end if

end if

end if

Out:

filtera - set filter for list A

filterb - set filter for list B

lista- set list for filter A

listb- set list for filter B

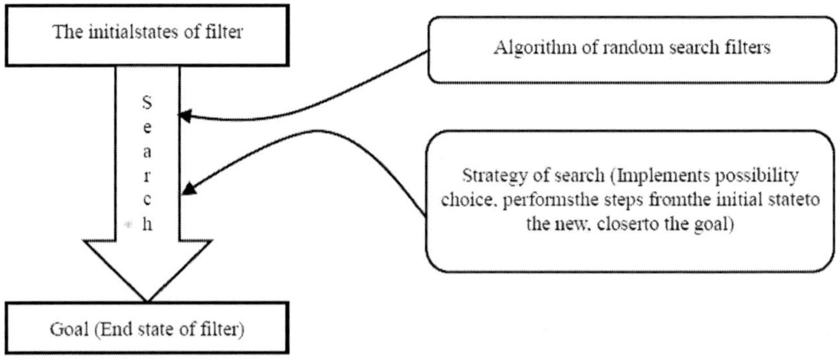

Figure 9. Structure procedure of random search filters.

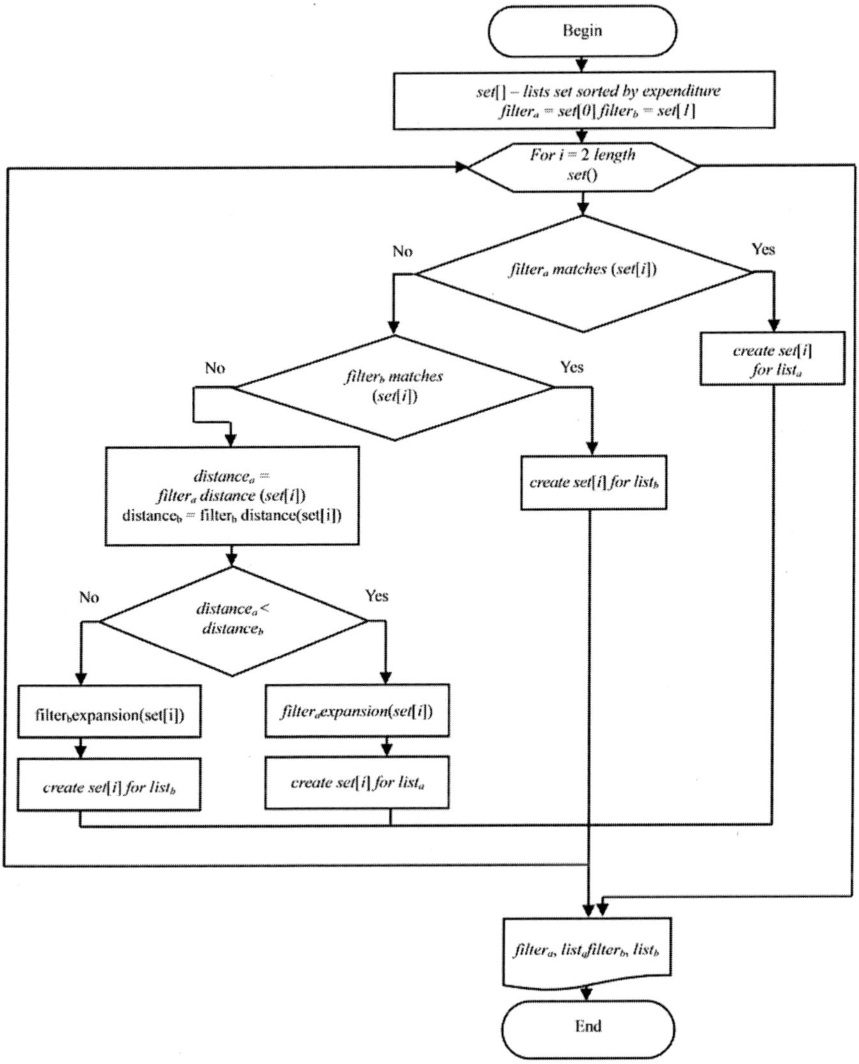

Figure 10. Flowchart algorithm ofa random searchfilters.

6. EVALUATE THE EFFICIENCY OF ACCELERATING PROCESS TRAFFIC FILTERING PROPOSED BY HAOTF

For evaluate the efficiency of accelerating process traffic filtering proposed by HAOTF is proposed scheme, which is shown in Figure 11.

There are two types of traffic and it is in the worst case condition is simulated traffic and traffic consists of the same packet that does not match any of the sets. This ensures that the package will be caught only the default action. Emulation traffic generated by creating packages that coincide with each set and the direction of their movement proportional to filter traffic. Traffic filtering action in the worst case is determined by measuring the size and permanent traffic CPU. Size is determined by the traffic load of traffic filtering use from 35% to 100%, the established list of set of rules.

6.1. Evaluate of Processing Attacks

The aim of this research is to test the strength of HAOTF the presence of attacks and traffic fluctuations. Since the number of hits set the default action is great and unpredictable, it can cause a great degree of vulnerability of all the work to filter traffic. To test the performance of HAOTF in handling such attacks are simulated attacks and increases algorithm of random search filters, is determined action set of rules of default from 0 to 200,000 (see Figure 12).

6.2. Evaluate of the Sensitivity Analysis

The final research was conducted for the sensitivity analysis of the proposed HAOTF. Analysis was performed for the set rules to different sizes from 0 to 5000 (see Figure 13).

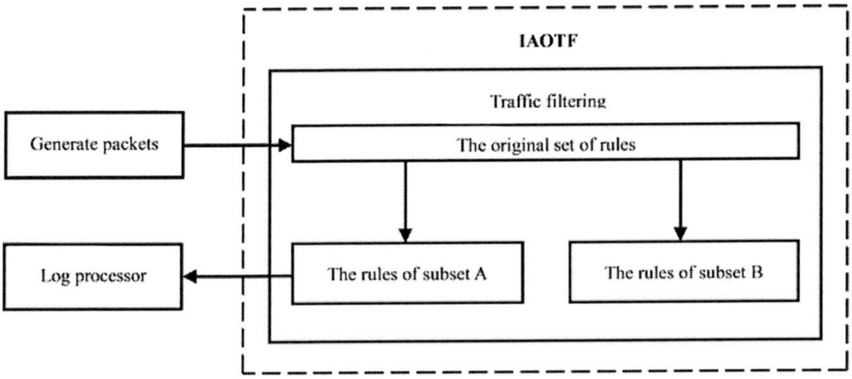

Figure 11. Scheme evaluate the efficiency of accelerating process traffic filtering proposed by HAOTF.

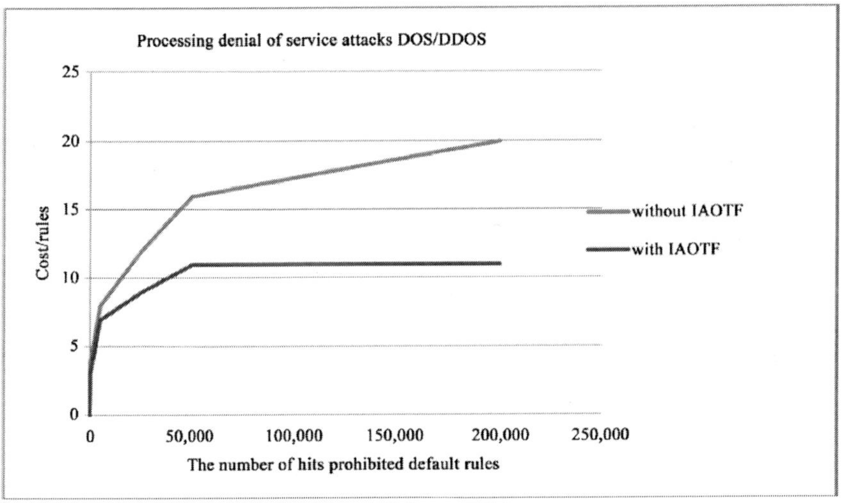

Figure 12. Confrontation DOS / DDOS attacks.

Figure 13 shows a comparative research between the list of set of rules traffic filtering, optimal, the algorithm of optimal search solutions and the algorithm of random search filters. The evaluation was conducted for heavily loaded traffic filtering process.

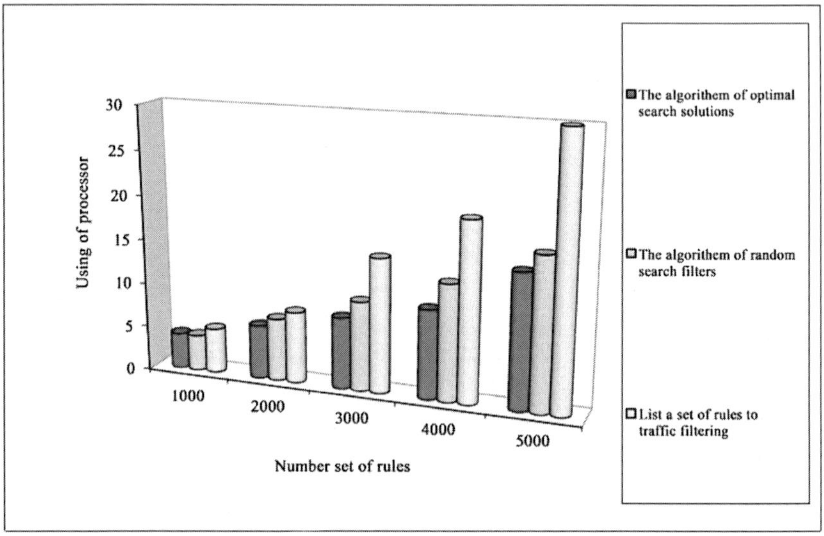

Figure 13. Sensitivity analysis.

7. CONCLUSION

This article devoted to the problem of optimization traffic filtering. For this, the article proposes a hierarchical structure for optimization traffic filtering. The basic principle of this structure is to provide lower overhead costs, semantic integrity and security of the initial set of rules to traffic filtering based on two algorithms, such as the algorithm of optimal search solutions and the algorithm of random search filters. In the processing of analysis of the results, it can be concluded that the proposed algorithm of random search filters is suited for hierarchical optimization of traffic filtering.

REFERENCES

1. Schneider, F., Wallerich, J. and Feldmann, A. (2007) Packet Capture in 10-Gigabit Ethernet Environments Using Contemporary Commodity Hardware. Springer, Berlin Heidelberg, 207-217.

2. Wu, W. (2011) G-NetMon: A GPU-Accelerated Network Performance Monitoring System for Large Scale Scientific Collaborations. In IEEE LCN'11, 2001.

3. Lu, G., Guo, C., Li, Y., Zhou, Z., Yuan, T., Wu, H., Xiong, Y., Gao, R. and Zhang, Y. (2011) Server Switch: A Programmable and High Performance Platform for Data Center Networks. Proceedings of the 8th USENIX Conference on Networked Systems Design and Implementation, Boston, 1 April 2011.

4. Gulomov, S., Abdurakhmanov, A. and Nasrullaev, N. (2015) Design Method and Monitoring Special Traffic Filtering under Developing «Electronic Government». International Journal of Emerging Technology & Advanced Engineering (ISSN 2250-2459, ISO 9001:2008 Certified Journal), Volume 5, Issue 1, January, 2015.

5. Karimov, M.M., Ganiev, A.A. and Gulomov, S.R. (2014) Models of Network Processes for Describing Operation of Network Protection Tools. 4th International Conference on Application of Information and Communication Technology and Statistics in Economy and Education, Sofia, 24-25 October 2014.

CHAPTER 2

Survey on Spam Filtering Techniques

Saadat Nazirova

Institute of Information Technology of Azerbaijan National Academy of Sciences, Baku, Azerbaijan

ABSTRACT

In the recent years spam became as a big problem of Internet and electronic communication. There developed a lot of techniques to fight them. In this paper the overview of existing e-mail spam filtering methods is given. The classification, evaluation, and comparison of traditional and learning-based methods are provided. Some personal anti-spam products are tested and compared. The statement for new approach in spam filtering technique is considered.

KEYWORDS

E-mail Spam, Unsolicited Bulk Messages, Filtering, Traditional Methods, Learning-Based Methods, Classification

1. INTRODUCTION

Spam is the use of electronic messaging systems (including most broadcast media, digital delivery systems) to send unsolicited bulk messages indiscriminately [1]. In this article it is considered the e-mail spam. E-mail spam, also known as junk e-mail or unsolicited bulk e-mail (UBE), is a subset of spam that involves nearly identical messages sent to numerous recipients by e-mail [2]. Day by day the amount of incoming spam increase and, scammer attacks are becoming targeted and consequently more of a threat. When targeted attacks first emerged five years ago, Symantec MessageLabs Intelligence tracked between one or two attacks per week. Subsequently, attacks have increased further from approximately 10 per day to approximately 60 per day in 2010 (**Figure 1**). By the end of 2010 MessageLabs Intelligence identified approximately 77 targeted attacks blocked each day [3].

By the Symantec MessageLab forecast spam will become more culturally and linguistically diverse, in 2011. The amount of spam sent from European countries will increase to 40% - 45% of all spam [3]. These facts state that spam is a big problem for today and for tomorrow and it actually makes sense to investigate new effective methods against spam.

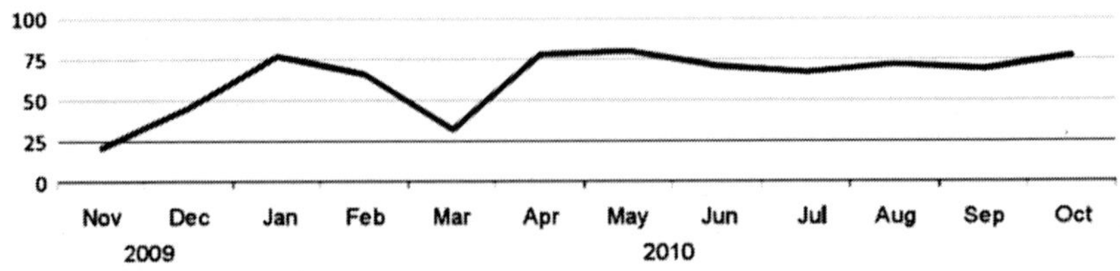

Figure 1. Targeted spam attacks 2009-2010 (Symantec. cloud Message Lab) [3].

2. HISTORICAL REVIEW OF SPAM FILTERING METHODS

Though the first spam was sent in 1978 it began to be written about it as a problem in scientific literature only from 1982. One of the first papers where this problem is considered is the Peter J. Denning's article [4]. The first mathematical apparatus applied to spam filtering systems is the Bayes' algorithm, which was used first by Sahami et.al in 1996 and then by other researchers [5-8]. Bayes' classifier relies on famous Bayes theorem and the first papers about it could be met as early as 1960 [9]. During more than 40 year history Naive Bayes Classifier (NBC) was used for the solution of very different type of tasks: from classification of texts in news agencies till primary diagnosis of diseases in medicine. For the problems where NBC is applied there is usually selected presence or absence of words in the text as a characteristic, i.e. the set of characteristics T is a set off all words in documents. Hereby, if the word t_i is present, the weight of characteristics $w_i = 1$, otherwise $w_i = 0$. In case of e-mail filters where spam classification is used, there taken into account the area where the word had been met: heading, subject and body of the e-mail.

Beginning from the publication of Gary Robinson [10], in some filters (for example, Spam Assassin) there came to be used the method of overlapping probabilities suggested by R. Fisher in 1950. For spam detection Robinson offered to calculate not only the probability of "spamness" of the document, but also the probability of "legitimness" of email. The next directions were application of Markov chain PageRank and Hidden Markov Model which are met in papers Paolo B., et al. [11], and José Gordillo, et al. [12]. Kolmogorov complexity estimation is met in papers Spracklin L.M., et al. [13]. Absolutely another approach is a new method of digital analysis of textual e-mails for spam detection which can be firstly observed in paper Korelov S. V., et al. [14]. Here e-mail is considered as a signal $x(n)$, after the methods of digital processing are applied to signals and the probability of false positives are defined for these methods.

Application of methods of clustering analyses to the problem of filtering e-mails to legitimate and spam is considered in papers [15-18]. From 2009 year, beginning from Paulo Cortez's, et al. article [19] one can meet the statement as a Symbiotic Data Mining which is a hybrid of Collaborative Filtering (CF) and ContentBased Filtering (CBF).

Considering stunning amount of spam messages coming to e-mail boxes it is possible to assume that spammers operate not alone, there are global, organized, virtual social networks of spammers. They attack e-mails of not only users, eve whole corporations and countries. Spam is of the weapons of information war. In spite of the fact that, the terms spam and war appear in one context [20,21] since 2003 year, only from 2009 the problem of spammers' social networks are considered in scientific papers. Clustering of spammers considering them in groups is offered in paper Fulu Li, et al. [22]. In works Xu K.S., et al. [23,24] the method of spectral clustering is applied to the set of spam messages collected under project Honey Pot for defining and tracking of social networks of spammers.

They represent a social network of spammers as a graph the nods of which correspond to spammers, and a corner between two junctions of graph as social relations between spammers.

Research and development of spam filtering systems are actively carried all over the world. Along with scientific institutes there are many organizations and corporations investigating and offering different theoretical, practical and juridical approaches to spam filtering. Different organizations as university laboratories (laboratories CSAIL MIT in USA [25], Computer Laboratory Faculty University of Cambridge in UK [26] and etc.); research centers (NCSR Democritos in Greece [27], research centre of IBM [28,29] and etc.); commercial companies (Microsoft [30], Symantec [31], Kaspersky's Laboratory [32] and etc.) had been involved to this process. Many international organizations take great attention to concerned problem. It is created the ASRG (AntiSpam Research

Group) [33] within the organization IETF (Internet Engineering Task Force) [34] in 2003. A lot of international conferences, summits and symposiums dedicated to this topic recently (NIST Spam Technology Workshop USA 2004, ASRG Meeting USA 2003, Cambridge Spam Conference USA 2003-2005, Conference on Email and Anti-Spam (CEAS) Mountain View USA 2004-2005, Spam Forum Paris France 2003, AntiSpam-Symposium Karlsruh Germany 2003, Spam Summit UK 2003, Conference on "Spam problem and its solution" Moscow and etc.) were held [35].

3. CLASSIFICATION OF SPAM-FILTERING METHODS

Depending on used techniques spam filtering methods are generally divided into two categories:

1) Methods to avoid spam distribution in their origins;

2) Methods to avoid spam at destination point.

Let's consider these methods in detailed form

3.1. Methods to Avoid Spam Distribution

Legislative measures limiting spam distribution, development of e-mail protocols using sender authentication, blocking mail servers which distribute spam are the methods which avoid spam distribution in origin.

Using these methods alone doesn't give considerable results. For example, there are many hard legislative restrictions for spam distribution in USA; nevertheless, the greatest amount of spam is distributed from this region. One of the reasons is an existence of high level broadband Internet access in USA. There is a number of the approaches, offering to make spam sending economically unprofitable. One of these statements is to make sending of each e-mail paid. The payment for one e-mail should be the extremely insignificant. In this case for the usual user it will be imperceptible. For spammers who send thousand and millions messages the cost of such mailing becomes considerable that makes it economically unprofitable.

This type of methods avoiding spam in their origins is a subject of author's another papers [36,37]. They should be implemented together with the methods described in the next section, which filter spam at the destination point.

3.2. Methods to Avoid Spam Receiving

Methods which filter spam in destination point can be divided into the following categories:

- Depending on used theoretical approaches: traditional, learning-based and hybrid methods;
- Depending on filtration area: server side, client side and filtration in public mail-servers.

3.2.1. Classification of Spam Filtering Methods Depending on Theoretical Approaches

As we noted above depending on used theoretical approaches spam filtering methods are divided into traditional, learning-based and hybrid methods. In traditional methods the classification model or the data (rights, patterns, keywords, lists of IP addresses of servers), based on which messages are classified, is defined by expert. The data storage collected by experts is called as the knowledge base. There are also used trusted and mistrusted senders lists, which help to select legal mail. Actually it makes sense only creation of the “white” list, because spammers use fictitious e-mail addresses. This technique can’t represent itself as a high-grade anti-spam filter, but can reduce considerably amount of false operations, being a part of e-mail filtration system based on other classification methods.

In learning-based methods the classification model is developed using Data Mining techniques. There are some problems from the point of view of data mining as changing of spam content with time, the proportion of spam to legitimate mail, insufficient amount of training data are characteristic for learning-based methods.

Traditional methods. Traditional methods are divided into the following categories:

1) Methods based on analysis of messages. The received e-mail is analyzed for specific signs of spam on the base of:

- formal signs;
- content using signature in updated database;
- content applying statistic methods based on Bayes theorem;
- content by means of use SURBL (Spam URL Realtime Block Lists) [38], when run search for located references in e-mail and their verification under base of SURBL. This method is effective if instead of advertisement, the reference of website with advertisement is located in e-mail.

2) Detectors of mass distribution. Their task is to detect distributions of similar e-mails to the bulk of users. The following methods are used for the detection:

- users' voting (Razor / Pyzor) [39,40];
- analysis of e-mails coming through mail system (DCC) [41];
- receipt of e-mail to the spam "trap" and its following analyses (implemented in Symantec Brightmail Anti-Spam) [42].

Independent from a way of bulk detection the idea of a method is that for spam filtration the calculated e-mail signature (the control sum) is used. For the methods based on detection of repetitions two vital issues are characteristic. The first is a spam "personification". This means that each spam e-mail has insignificant differences at the cost of which it is hard to collect steady signatures. To solve this problem the various steady signatures are used. For example, in Yandex Mail System the method of shingles [43] is realized. The second problem is a detection of legitimate bulk mailings.

3) Methods based on acceptance of sender as a spammer. These methods relies on different blackhole lists of IP and e-mail addresses. It is possible to apply own blackhole and white lists or to use RBL services (Real-time Blackhole List) and DNSBL (DNS-based Blackhole List) for address verification. Advantage of these methods is detection of spam in early step of mail receiving process. Disadvantage is that the policy of addition and deletion of addresses is not always transparent. Often the whole subnets belonging to providers get to the Black lists. For such systems it is actually impossible to estimate the level of false positives (the legitimate e-mail wrongly classified as spam) on real mail streams.

4) Methods based on verification of sender's e-mail address and domain name. This is the simplest method of filtration if DNS request's name is the same with the domain name of sender. But spammers can use real addresses, so that current method is ineffective. In this case it may be verified with possibility of sending the message from current IP address. Firstly, the Sender ID technology [44] can be used where sender's e-mail address is protected from falsification by means of publishing the policy of

domain name use in DNS. Secondly, there can be used SPF (Sender Policy Framework) technology [45], where DNS protocol is used for verification of sender's e-mail address. The principle is that if domain's owner wants support SPF verification, then he adds special entry to DNS entry of his domain, where indicates the release of SPF and ranges of IP addresses from where may become an email from users of current domain.

5) Method based on SMTP server response emulation. If the real mail delivery systems, which follow the SMTP protocol correctly, observe such error, they get some interval (1 - 2 hours) and repeat attempt again [46]. But the majority of spam-bots has very short time out periods. So filters based on this method slow down the SMTP transaction to the point that some SPAM senders will fail but where real mail delivery systems will still continue and deliver mail successfully.

All above methods are based on some data for analysis collected by experts of third-party suppliers and same for all users. So that traditional method's has the following disadvantages:

- it is necessary to update the knowledge base regularly;
- there is a dependence on update suppliers;
- the security level is low;
- "impersonalized" model of classification doesn't consider individual specifics of user's correspondence;
- dependence on natural language of correspondence;
- low level of detection because of general models of classification.

Learning-based methods. Nowadays there is actively developed trainable or intellectual methods based on Data Mining algorithms for e-mail filtration. These algorithms divide the object to some categories using classification model previously defined on the base precedential information.

Assume spam filtration is defined by the function

$$f(m,\zeta)= \begin{cases} m_{spam}, & \text{if the message } m \text{ is considered as spam} \\ m_{leg}, & \text{if the message } m \text{ is considered as legitimate mail} \end{cases}$$

where m is a classified mail, ζ is a vector of parameters m_{spam} and m_{leg} are spam and legitimate e-mail.

Many spam filters based on classification using machine learning techniques. In learning-based methods the vector of parameters ζ is a result of classification trainings on previously collected e-mails.

$$\zeta = Z(M).$$
$$M = \{(m_1, y_1), (m_2, y_2), \cdots, (m_n, y_n)\},$$
$$y_i \in \{m_{spam}, m_{leg}\},$$

where $m_1, m_2, \cdots, m_n$ are previously collected messages, $y_1, y_2, \cdots, y_n$ are the corresponding labels and Z is the training function.

The following types are belonged to learning-based methods.

1) Image-based spam filtering. Image spam has become a new type of e-mail spam. Spammers embed the message into the image and then attach it to the mail. Some traditional methods based on analysis of text-based information do not work in this case. Image filtering process is costly and time-consuming work. In the paper [47] it is proposed three-layer (Mail Header Classifier, the Image Header Classifier and the Visual Feature Classifier) image-spam filtering. In the First layer it is applied Bayesian classifier and SVM classifier in the remaining layers. In paper [48] it is offered statistical feature extraction for classification of image-based

spam using artificial neural networks. They consider statistical image feature histogram and mean value of block of image for image classification.

2) Bag of words Model. The bag-of-words model is a simplifying assumption used in natural language processing and information retrieval. In this model, a text (such as a sentence or a document) is represented as an unordered collection of words, disregarding grammar and even word order [49]. In spam filtering two bags of words are considered. One bag is filled with word found in spam e-mails, and the other bag is filled with words met in legitimate e-mails. Considering e-mail as a pile of words from one of these bags, there used Bayesian probability to determine to which bag this e-mail belongs. k-Nearest neighbor, SVM (Support Vector Mashine), boosting classifiers are also applicable to the bag of words.

3) Collaborative spam filtering. This is gathering spam reports between P2P users or from mail server (Google Gmail). The collaborative centralized spam filtration is more economic in comparison with personal approach, but only under condition of presence of adequate procedures of the analysis of false operations and operative reclassification of not correctly classified messages. In the papers [50-54] it is proposed such kind of multi-agent spam filtration and personalized collaborative spam filtering.

4) Social networking against spam. This is a one of the latest methods where the information extracted from social networks is used to fight spammers. For example, P.A. Chirita et al. [55] estimate the rank of users depending on their social network activities and trustworthy senders are ranked and classified as spam or non-spam. They call this algorithm as MailRank schema and show that it is highly resistant against spammer attacks, which obviously have to be considered right from the beginning in such an application scenario.

So in case of learning-based methods user defines the classification model himself, so that the majority disadvantages of traditional methods are solved successfully; intellectual methods are autonomous, independent on external knowledge base, doesn't require regular update, multilingual, independent of natural language, able to study new types of spam user-aided. There is advantage as construction of personalized mail classification model, where user himself defines which mail is legal or which one is a spam. Therefore learning-based methods have higher rank in spam determination. In many spam filtration systems based on the learning-based methods the Bayes' theorem, Marcov's chain and others are successfully applied. Learning-based methods have also a couple of disadvantages as overfitting, dependence on quality and compound of trainee set, resource-intensivety. Application of statistic algorithms with complicated mathematic calculations led to high loading of computing system's resources. For the spam filtering systems processing fair amount of requests the productivity of algorithm is a main importance, so resource-intensivety factor is the most important disadvantage of learningbased methods.

Hybrid methods. One of the latest approaches in spam filtering is hybrid filtration system which is a combination of different algorithms, especially if they use unrelated features to produce a solution. In this case it can be applied various filtering techniques and get the advantages of the traditional and learning-based methods [56].

3.2.2. Classification of Spam Filtering Methods Depending on Filtration Scope

Depending on filtration scope spam filtration methods are divided into the following categories.

1) Client side/personal filters. Client side filters works directly on user's computer. In client side filtration e-mail loading to the user's local computer anyway, and only after that classified what

leads to additional loading of data transfer in network. Client side spam filtration more accurately due to usage methods of machine learning. In client side filters users' personal information are used, in server side filters the filtration model is defined at once for all users. In spite of the fact that for the majority of users it is obvious what is spam, the concept of spam for each of them is enough personified. The e-mail message marked as spam by someone may be the important information for other one. From filtration quality point of view the personal model is the most preferable as characteristics of user's correspondence are considered. Generally, absence of personification reduces the level of detection and increases quantity of false positives. On the other hand, use of personal model of e-mail classification involves an inevitable overhead cost. Firstly the user should construct his personal model of filtration himself as only he can define what legal e-mail is, and what spam is for him. Secondly, construction, storage and use of personal model demands additional computing resources.

2) Server side/general filters. Server side filters work at mail server level. Generally in server side filtration systems the traditional methods of filtration are applied, but at client level the learning-based or hybrid one. Server side filtration also own priority. As centralized solution reduces expenses, simplifies support and control of this system. User becomes more mobile, so that it is comfortable to store mail centralized in server and to have an access to him from different points, using different devices. Hereby, classification at mail-server level more preferably and development of these methods more actual.

3) Spam filtering in public mail-servers. This solution sometimes is better than client or server solution. In this case users are mobile as in case of server side filtration, and personalized as in case of client side solution. But disadvantage of usage of public mail-servers is that users depend on filtration product installed there. For example, the mail-server of Google.Inc company gmail. com

uses its own products against spam [57]. This system considers personal information about user to minimize false positives. The public mail provider Mail.ru uses Kaspersky Anti-Spam product based on "Spamtest" technology, and absolutely based on traditional filtration methods, as well RBL, the base of fuzzy signature of mails with spam, heuristics base. These knowledge bases are maintenanced and updated regularly till 3 times in an hour. Processing of attached files, detection of iterations is supported also. The system as a general model of classification applicable for all users, but at the same time personalization is absent.

4. SOFTWARE SOLUTIONS FOR SPAM FILTERING

On basis of suggested theoretical approaches different companies develop hardware software solutions for spam filtering. In this paper it is given the result of testing of personal spam filters. Choosing anti-spam software it is necessary to compare:

- price;
- functionality and user-friendness;
- quality of spam detection.

Estimation of cost and functionality can be made by the company-manufacturer documentation. The quality of spam detection can be defined by the testing these products [58].

In this paper there seven freeware anti-spam software products are chosen for testing. Each soft is installed on Windows XP platform on different personal computers with POP3 mail server. The feature of our test environment was that for tested products the real post traffic was used.

Testing was made 14 days, and during this period 721 messages has been processed, 430 from which were spam. Learning-based filters were previously trained with 33 spam and 33 legitimate e-mails. The spam classification result of each software is represented in **Table 1**. The results of filtration were divided into four categories: false positive, true positive, true negative, false negative. The following parameters were calculated false positive percent (FPP)—the percent of legitimate e-mails detected as spam and the parameter false negative percent (FNP)—the percent of not detected spam e-mails.

It is better to get spam than to lose the legitimate e-mail with important information, so the low level of FPP is important than low level of FNP. The result of experiment shows that the lower values of FNP were in filters Qurb 3.0, Matador 1.0.0 and SpamArrow (**Figure 2**).

Table 1. Testing of different spam filtering software products.

Anti-spam software	FPP	FNP	FP	Detected		FN
				TP	TN	
Matador 1.0.0	4.9%	4.7%	35	410	256	256
SpamBrave	1.1%	7.0%	8	400	283	283
SpamArrow	2.5%	5.8%	18	405	273	273
Espresso 1.06.94	2.9%	20.9%	21	340	270	270
Spam Bully	2.1%	16.3%	15	360	276	276
Spam Fighter	1.9%	19.1%	14	348	277	277
Qurb 2.0	5.8%	2.3%	42	420	249	249

TH-True Negatives, TP-True Positives, FN-False Negatives, FP-False Positives.

And for the FPP the lower value were in filters Spam Brave, Spam Fighter and SpamArrow. The best effective between considered spam filtering software products is SpamArrow. Its efficiency is connected by that it is learning-based and it was trained previously. So for effective spam filtering it is necessary to install traditional filter before learning-based one to collect spam templates for previous training.

5. CONCLUSIONS

Summarizing above-listed, we obtain the following conclusions.

So, spammers constantly change external signs of e-mails to skip spam filtering systems, there arises a need for adaptive filtering system, which should have the ability to react quickly to the changes and provide fast and qualitative self-tuning in accordance with a new set of features.

Since the filters are trained on a very limited number of messages that come only to a specific user or a specific mail provider, the quality of filtration in the existing client and server filtering systems is rather low. But it can be improved if to apply the hybrid filtration system in other words the complex hierarchical and multi-agent filtration system that helps users to participate in the identification of the filtering errors and the appropriate setting of filters at each level (user level, organization level, mail provider level).

Therefore it is quite perspective for solving this problem, the combination of two widespread approaches as using the personal e-mail classification model on a server side solution. Development of server side personalized e-mail filtering systems that use the learning-based classification algorithms based on Data Mining methods is a very perspective direction.

This statement is supported by the followings:

- personalized server side filtering systems are preferable than the client side solutions, because provide universal access to an e-mail, reduce expenses, which is very important for corporate users;

- personalized server side filtering systems are more preferably because of greater accuracy and fewer errors in comparison with general model;

- personalized server side filtering system offered in author's another paper [59] bases on the Universal Declaration of Human Rights and has a universal character, can be applied in all countries;

learning-based algorithms used in personalized server side filtering systems exceed traditional ones because of a number of fundamental qualities (quality of filtering, the absence of updates, autonomy, independence from external knowledge bases).

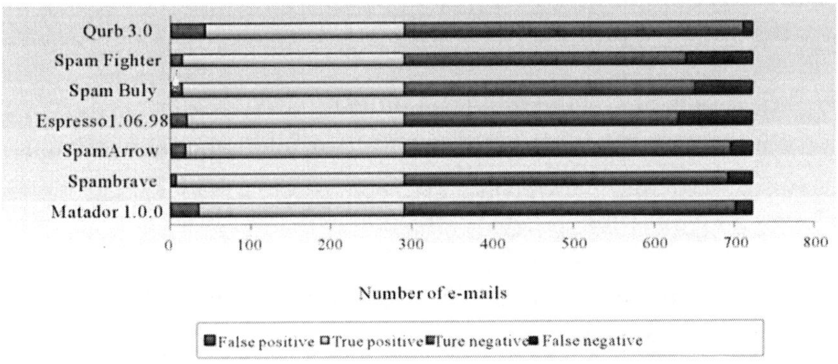

Figure 2. Spam filtering software products testing diagram.

ACKNOWLEDGEMENTS

I would like to give my sincere thanks to Dr. Rasim Alguliev and Dr. Ramiz Aliguliyev for their ideas.

REFERENCES

1. Wikipedia, “Spam”. http://en.wikipedia.org/wiki/ Spam_ (electronic)

2. Wikipedia, “E-mail spam”. http://en.wikipedia.org/wiki/E-mail_spam

3. Symantec, “State of Spam and Phishing. A Monthly Report 2010,” 2010. http://symantec.com/content/en/us/enterprise/ other_rsources/b-state_of_spam_and_phishing_report_09-2010.en-us.pdf.

4. J. P. Denning, “ACM President’s Letter: Electronic Junk,” Communications of the ACM, Vol. 25, No. 3, March 1982, pp. 163-165. doi:10.1145/358453.358454

5. M. Sahami, “Learning Limited Dependence Bayesian Classifiers,” Proceedings of the Second International Conference on Knowledge Discovery and Data Mining, The AAAI Press, Menlo Park, 1996, pp. 334-338.

6. M. Sahami, S. Dumais, D. Heckerman and E. Horvitz, “A Bayesian Approach to Filtering Junk Email,” AAAI Technical Report WS-98-05, AAAI Workshop on Learning for Text Categorization, 1998.

7. J. R. Hall, “How to Avoid Unwanted Email,” Communications of the ACM, Vol. 41, No. 3, 1998, pp. 88-95.

8. E. Gabber, M. Jakobsson, Y. Matias and A.J. Mayer, "Curbing Junk E-Mail via Secure Classification," Proceedings of the Second International Conference on Financial Cryptography, Springer-Verlag London, 23-25 March 1998, pp. 198-213.

9. R. A. Fisher, "On Some Extensions of Bayesian Inference Proposed by Mr. Lindley," Journal of the Royal Statistical Society: Series B, Vol. 22, No. 2, 1960, pp. 299-301.

10. G. Robinson, "A Statistical Approach to the Spam Problem," 2003. http://www.linuxjournal.com/article.php?sid=6467 (accessed March 2011).

11. P. Boldi, M. Santini and S. Vigna, "PageRank as a Function of the Damping Factor," Proceedings of the 14th International Conference on World Wide Web, ACM New York, 10-14 May 2005.

12. J. Gordillo and E. Conde, "An HMM for Detecting Spam Mail," Expert Systems with Applications, Vol. 33, No. 3, 2007, pp. 667-682.

13. L. M. Spracklin and L. V. Saxton, "Filtering Spam Using Kolmogorov Complexity Estimates," in Russian, 21st International Conference on Advanced Information Networking and Applications Workshops (Ainaw'07), Niagara Falls, 21-23 May 2007, pp. 321-328.

14. S. V. Korelov, A. K. Kryukov and L. U. Rotkov, "Text Messages' Digital Analysis on Spam Identification," in Russian, Proceedings of Scientific Conference on Radiophysics, Nizhni Novgorod State University, Nizhny Novgorod Oblast, 2006.

15. W.-F. Hsiao and T.-M. Chang, "An Incremental Cluster-Based Approach to Spam Filtering," Expert Systems with Applications, No. 34, No. 3, 2008, pp. 1599-1608.

16. S. M. Lee, D. S. Kim and J. S. Park, "Spam Detection Using Feature Selection and Parameters Optimization," IEEE International Conference on Intelligent and Software Intensive Systems, Krakow, 15-18 February 2010, pp. 883-888.

17. M. F. Saeddian and H. Beigy, "Spam Detection Using Dynamic Weighted Voting Based on Clustering," Proceedings of the 2008 Second International Symposium on Intelligent Information Technology Application, Vol. 2, pp. 122-126.

18. M. Sasaki and H. Shinnou, "Spam Detection Using Text Clustering," IEEE Proceedings of the 2005 International Conference on Cyberwords, Singapore, 23-25 November 2005, pp. 316-319.

19. P. Cortez, C. Lopes, P. Sousa, M. Rocha and M. Rio, "Symbiotic Data Mining for Personalized Spam Filtering," IEEE/WIC/ACM International Conference on Web Intelligence and Intelligent Agent Technology, Milan, 15-18 September 2009, pp. 149-156.

20. W. Lauren, "Spam Wars," Communications of the ACM — Program Compaction, Vol. 46, No. 8, 2003, p. 136.

21. G. Pawel and M. Jacek, "Fighting the Spam Wars: A Re-Mailer Approach with Restrictive Aliasing," ACM Transactions on Internet Technology (TOIT), Vol. 4, No. 1, 2004, pp. 1-30.

22. F. Li, H. Mo-Han and G. Pawel, "The Community Behavior of Spammers" 2011. http://web.media.mit.edu/~fulu/ Clustering Spammers.pdf.

23. K. S. Xu, M. Kliger, Y. Chen, P. J. Woolf and A. O. Hero, "Revealing Social Networks of Spammers through Spectral Clustering," IEEE International Conference on Communications, Dresden, 14-18 June 2009, pp. 1-6.

24. K. S. Xu, M. Kliger and A. O. Hero, “Tracking Communities of Spammers by Evolutionary Clustering,” 2011. http://www.eecs.umich.edu/~xukevin/xu_spam_icml_2010_sna.pdf.

25. Laboratory CSAIL MIT in USA, 2011. http://projects. csail.mit.edu/spamconf/.

26. Computer Laboratory Faculty Cambridge University in UK, 2011. http://www.cl.cam.ac.uk/~rnc1/.

27. National Center for Scientific Research, “Demokritos,” 2011. http://www.iit.demokritos.gr/.

28. D. Mertz, “Spam Filtering Techniques,” 2002. http://www.ibm.com/developerworks/linux/library/l-spamf.html.

29. R. Segal, J. Crawford, J. Kephart and B. Leib, “SpamGuru: An Enterprise Anti-Spam Filtering System,” IBM Thomas J. Watson Research Center. http://www.research.ibm.com/ people/r/ rsegal/papers/spamguru-overview.pdf.

30. Microsoft Antispam Technologies. http://www.microsoft.com/mscorp/safety/technologies/antispam/default.mspx.

31. Symantec Antispam Protection for E-Mail. http://www. symantec.com/business/premium-antispam,.

32. Kasperskiy Ant-Spam. http://www.kaspersky.ru/anti-spam.

33. Anti-Spam Research Group. http://asrg.sp.am/.

34. The Internet Engineering Task Force. http://www.ietf.org/.

35. Spam Events. http://spamlinks.net/conf.htm.

36. S. A. Nazirova, “Anti-Spam Module for Filtering the Outgoing Correspondence,” in Russian, Transactions of ANAS, Informatics and Control Problems, Vol. XXVIII, No. 3, 2008, pp. 158-162.

37. S. A. Nazirova, "New Anti Spam Methods," Proceedings on the Second International Conference on Problems of Cybernetics and Informatics, Baku, 10-12 September 2008, pp. 89-92.

38. Spam URL Realtime Block Lists. http://www.surbl.org/.

39. Razor's homepage. http://razor.sourceforge.net/.

40. Pyzor's homepage. http://sourceforge.net/apps/trac/pyzor/.

41. DCC Spam Control Delayed Your E-Mail. http://mail.cc.umanitoba.ca/grey/.

42. Symantec Brightmail Anti-Spam. http://www.symantec.com/business/premium-antispam.

43. Yandex, "Some Automatic Spam Detection Methods". http://company.yandex.ru/public/articles/antispam.xml.

44. Microsoft Sender ID Framework. http://www.microsoft.com/mscorp/safety/technologies/senderid/default.mspx.

45. Sender Policy Framework. http://www.openspf.org/Introduction.

46. J. Klensin, "RFC-2821: Simple Mail Transfer Protocol," April 2001. http://www.rfc-ref.org/RFC-TEXTS/2821/index.html.

47. T.-J. Liu, W.-L. Tsao and C.-L. Lee, "A High Performance Image-Spam Filtering System," Ninth International Symposium on Distributed Computing and Applications to Business, Engineering and Science, 10-12 August 2010, Hong Kong, pp. 445-449.

48. M. Soranamageswari and C. Meena, "Statistical Feature Extraction for Classification of Image Spam Using Artificial Neural Networks," Second International Conference on Machine Learning and Computing, Bangalore, 9-11 February, 2010, pp. 101-105.

49. Bag of Words Model. http://en.wikipedia.org/wiki/ Bag_of_words_model_in_computer_vision.

50. K. Li, Z. Zhong and L. Ramaswamy, "Privacy-Aware Collaborative Spam Filtering," IEEE Transactions on Parallel and Distributed Systems, Vol. 20, No. 5, May 2009, pp. 725-739.

51. F. Weidong and D. Shoubin, "Addressing Interest Diversity in P2P Based Collaborative Spam Filtering," Fifth International Conference on Grid and Cooperative Computing Workshops, Hunan, October 2006, pp. 163-169.

52. J. S. Kong, B. A. Rezaei, N. Sarshar, V. P. Roychowdhury and P. O. Boykin, "Collaborative Spam Filtering Using E-Mail Networks," IEEE Computer Society on Computer, Vol. 39, No. 8, 2006, pp. 67-73.

53. A. Gray and M. Haahr, "Personalised, Collaborative Spam Filtering," Proceedings of the First Conference on Email and Anti-Spam (CEAS), Mountain View, 30-31 July 2004.

54. R. M. Alguliyev and S. H. Nazirova, "Multilayer and Multiagent Automated Email Filtration System," Telecommunications and Radioengeneering, Vol. 67, No. 12, pp. 1089-1095.

55. P. A. Chirita, J. Diederich and W. Nejdl, "MailRank: Using Ranking for Spam Detection," Proceedings of the 14th ACM International Conference on Information and Knowledge Management, Bremen, 31 October-5 November 2005.

56. R. Bhuleskar, A. Sherlekar and A. Pandit, "Hybrid Spam E-Mail Filtering," 2009 First International Conference on Computational Intelligence, Communication Systems and Networks, Indore, 23-25 July 2009, pp. 302-307. doi:10.1109/CICSYN.2009.34

57. Google Message Security Postini Services. http://www.google.com/postini/email.html.

58. R. M. Alguliyev and S. H. Nazirova, “Architecture of Hierarchical Intellectual Nation-Wide System of Struggle against Spam,” in Russian, Information Technologies, Moscow, No. 8, 2006, pp. 32-36.

59. R. M. Alguliyev and S. H. Nazirova, “Mechanism of Formation and Realisation of Anti-Spam Policy,” in Russian, Telecommunications, Moscow, No. 12, 2009, pp. 6-10.

CHAPTER 3

Efficient Spam Filtering System Based on Smart Cooperative Subjective and Objective Methods*

Samir A. Elsagheer Mohamed[1,2]

[1] College of Computer, Qassim University, Qassim, KSA

[2] Electrical Engineering Department, Faculty of Engineering, Aswan University, Aswan, Egypt

ABSTRACT

Most of the spam filtering techniques are based on objective methods such as the content filtering and DNS/reverse DNS checks. Recently, some cooperative subjective spam filtering techniques are proposed. Objective methods suffer from the false positive and false negative classification. Objective methods based on the content filtering are time consuming and resource demanding. They are inaccurate and require continuous update to cope with newly invented spammer's tricks. On the other side, the existing subjective proposals have some drawbacks like the attacks from malicious users that make them unreliable and the privacy. In this paper, we propose an efficient spam filtering system that is based on a smart cooperative subjective technique for content filtering in addition to the fastest and the most reliable non-content-based objective methods. The system combines several applications. The first is a web-based system that we have developed based on the proposed technique. A server application having extra features suitable for the enterprises and closed work groups is a second part of the system. Another part is a set of standard web services that allow any existing email server or email client to interact with the system. It allows the email servers to query the system for email filtering. They can also allow the users via the mail user agents to participate in the subjective spam filtering problem.

KEYWORDS

Anti-Spam System; Objective Spam Filtering; Cooperative Subjective Spam Filtering; Web Application Web Services

1. INTRODUCTION

With the rapid growth of the Internet, email has become one of the most common media for us to exchange information. More and more people depend on it to communicate. Spam, unsolicited, undesired, bulk email (or junk email), has been a significant security issue for computer users and a massive waste of time, disk spaces, and network bandwidths. Spam has been recognized as problem since 1975 [1]. Spam is usually characterized as unsolicited commercial or bulk email, is sent in large numbers and repeatedly to individuals. According to the statistics from ITU (International Telecommunication Union), 70% to 80% of the present emails in Internet are spam, which has become a worldly problem to the information infrastructure [2,3].

Spam is a multifaceted phenomenon and therefore very complicated to address. This phenomenon is probably one of the biggest challenges the Internet will have to face in the immediate future [4]. A spammer sends a large number of messages to many different recipients who have not requested the content. (Interestingly most spammers do not care whether a particular addressee receives the message; they merely seek to get a sufficient percent of their postings delivered to some of the addressees.) Spam can conform to Internet technical standards and can contain no technical differences from legitimate—desired—messages.

Moreover, the association of spammers with hackers and virus writers poses a very real threat to the Internet security and availability. About 8 years ago, spam was sent by spammer's own e-mail servers. Approximately 45% - 60% of spam is now sent from compromised systems distributed over the Internet [5-8]. Spam relaying increases the distribution base and at the same time eludes and overwhelms spam detection systems [9].

To fight the spam, there exist many objective measures that can effectively limit the spam email impact on the end users (subjects). Traditional anti-spam techniques include the Bayesian-based filters [10-13], rule-based Scoring Systems [14-16], DNS MX Record Lookup and Reverse Lookup Systems [17], DNS Realtime Blackhole List (DNSRBLs) or IP Blacklists [18]. However, these objective measures fail in many situations and they can have a level of accuracy. They suffer from the false positive and the true negative problems. Where a non-spam email can be classified as spam email or the filter can classify a spam email as legitimated email (non-spam). The spam problem is very complex one. An ideal anti-spam solution cannot be found as the spammers constantly try to invent new tricks and mechanisms to bypass the objective anti-spam measures so that their spam emails be classified as legitimated emails.

Emails are sent to human (subjects). Thus, only humans can accurately identify spam emails based on the content. As a result, subjective methods are better than the objective methods for content-based filtering. If we can have a group of qualified users (subjects) to monitor and read all the incoming emails, sure this will be the ideal solution for spam filtering. However, for many reasons this solution cannot be applicable because of the privacy, the scalability, the huge volume of emails that is sent by the spammers, etc.

As we highlighted, objective measures alone cannot fight the spam problem. In addition, no pure subjective measures can be implemented effectively to work in real time. Subjective measures can check only the message content (the email body), but in no means they can check other spamming metrics like a forged sender email. Objective measures can accurately and rapidly eliminate in real time up to 70% of spam emails based on known spamming techniques. The only time-consuming and inaccurate objective measures are those filtering the spam based on the body contents.

In this Paper, we propose a new technique for the spam filtering problem. It consists in combining both subjective and objective methods. This can effectively be accurate and reliable spam filtering solution. If subjective measure is devised in such a way to maintain the privacy of the users, it can cope with almost all the newly devised spammer's tricks that can pass through the objective measures. In addition, they can compensate (correct) the false-positive objective evaluations.

In addition, we have developed a complete system that is based on the proposed technique. The system consists of several software components. The first is a web-based email client application that allows users to access their email accounts from a central location. It allows them to benefit from combining both the traditional objective spam filtering and the smart subjective spam filtering. Users are the main players in the subjective spam filtering, thus through this application, they can do that in a smart way. Another application is also presented that can be used for enterprises and companies to allow them to add all their users to the systems and make them benefit from the offered solutions as well as participate in the subjective spam filtering problem. A web service is developed that allows any email server to request from the system the subjective evaluation of the newly arrived emails. Another developed web services allow the users of any email user agent to participate in the subjective spam filtering. All these open source software are presented in the paper.

The rest of this paper is organized as follows: In Section 2, the related works and research efforts are given. The description of the proposed technique is provided in Section 3. The developed Email Client Application that is using the proposed technique is described in Section 4. In addition, the cooperative subjective application is described in Section 5. In Section 6, we provided a set of developed web services that allows any other email application to interact with the developed system. Finally, the conclusions and the future works are given in Section 7.

2. RELATED WORKS

Various techniques [3,19-21] on automatically detecting or filtering spam emails have been proposed by building comprehensive databases for blocking emails whose addresses have been reported as black-lists or whose message bodies contain specific words or phrases defined as threatening terms, etc. The most famous approach based on the statistical filtering is the Bayesian filter [11-13], which is based on the Bayes' theorem. This theorem states that the probability that an email is spam, given that it has certain words in it, is equal to the probability of finding those certain words in spam email, times the probability that any email is spam, divided by the probability of finding those words in any email. The drawback of the statistical filtering techniques is the processing time: the time required to process an email and to end up if it is a spam or not. Another problem with this kind of filtering is that it cannot fight against the new tricks of the spammers, like changing vocabulary, introducing the most recognizable terms or adding a relatively high number of random words, miss spell words, adding numbers and symbols in the middle of the word or the phrase, etc.

Another new direction for the spam filtering problem which becomes now more popular is the Collaborative filtering [11,22] approach, which is a generic approach used to describe any type of filtering in which a network of intelligence is used to identify spam. A collaborative intelligence network can take many forms, such as a collection of lexical data in which characteristics of spam are described. In most cases, collaborative networks take advantage of the misfortune of others receiving spam to build better intelligence for future filtering efforts. This idea is implemented in Vipul's Razor [23], which is a distributed, collaborative, spam detection and filtering network. Through user contribution, Razor establishes a distributed and constantly updating catalogue of spam in propagation that is consulted by email clients to filter out known spam. Detection is done with statistical and randomized signatures that efficiently spot mutating spam content. The drawback of collaborative filtering is the participating community itself [11,22]. In large communities, there is either a high maintenance loop or, if automated, a high risk of false positives. Larger networks with high maintenance loops generally experience latency in updating their databases, in the same way that blacklisting has a propagation delay. Automated networks run the risk of false information being propagated or possibly injected by a malicious party. Smaller networks are generally more accurate and more real-time but lack the ability to cover a wide pool of fresh inbound spam.

The existing anti-spam techniques are passive, that is to wait for the spam email and to do all the best to identify the spam email and move it to the Spam folder of the user or simply drop it. However, in [24], a completely different approach is proposed for the spam problem. An active technique that attack the spammers' resources. By grapping all the links from an incoming spam email and building an application that periodically download the contents from these links, the spammers resources (their servers and network bandwidth) will be wasted answering the request from this application. Once this application is widely deployed and used by many users, it will act as a massive Distributed Denial of Service (DDoS) attack on the spammers' resources. For more details see [24].

Another direction to fight the spam problem is the "machine learning" techniques. Machine learning is the ability of a machine to improve its performance based on previous results. The key feature of machine learning is that based on existing dataset training has to take place in order to learn to distinguish between spam and non spam emails. Feature extraction is a major and critical part of machine learning. It does not work on the raw content of the email, but extracting the features then feed them to the tool that can tell wither this email is spam or not based on the experience it has during the training or the learning phase. Moreover, the dataset has to be divided into three parts, one of the training, another for testing and the third for the cross validation. Once operational, it can give false positive or false negative. These false predictions can be recycled to dataset to improve the accuracy. Of course, a retraining algorithm and update criteria have to be done regularly to deal with the new tricks of the spammers.

Some of the existing machine learning based techniques are: rule learning, decision trees [25], support vector machines [26,27] or combinations of different learners [14,28]. The basic and common concept of these approaches is that using a classifier to filter out spam and the classifier is learned from training data rather than constructed by hand.

In [29], ant colony optimization (ACO) algorithm is proposed to detect spam in host level. In this work, the host link structure is first constructed by aggregating hyperlinks over web pages. Training and validation is required by ACO to distinguish between good (non spam) and bad (spam) web pages. The authors in [10] propose a social network Aided Personalized (SOAP) as a spam filter. SOAP does not focus on parsing key-words (e.g, Bayesian filter) or building blacklists, it exploits the social relationship among email correspondents to detect the spam adaptively and automatically. From the machine learning viewpoint, spam filtering based on the textual content of e-mail can be viewed as a special case of text categorization, with the categories being spam or non-spam [14,15,30].

Spam can also be sent embedded in an image [31,32] or in a PDF documents (i.e. the text of the message is converted to an image and sent by email or MMS to users). This kind of spam is very hard to detect as the spam detecting software has to decode the image and convert it back to text using Optical Character Recognitions (OCR) to identify if the content is a real spam or just a normal image. Only the human can read the text on the image. This trick of the spammers is one of the hardest to solve. Many research efforts concentrate to image spam detection as the ones given in [33,34]. In [33], a framework for filtering image spams by using the Fourier-Mellin invariant features is presented. Fourier-Mellin features are robust for most kinds of image spam variations. A one-class classifier, the support vector data description (SVDD), is exploited to model the boundary of image spam class in the feature space without using information of legitimate emails. However, these approaches have many limitations that limit their applicability in real time. They require too much computational power to process all the pixels in the image, usual many iterations have to executed with complex algorithms. Spammers use several techniques to make the text recognition almost impossible. They are language dependant techniques. There is no generic OCR for all the languages. This make these approaches ineffective for realistic email applications to run in real time due to the huge volume of emails coming to the server and the excessive processing complexity for each image embedded in an email.

3. THE PROPOSED SPAM FILTERING TECHNIQUE

As stated in the Introduction, it is known that subjective measures are better than the objective measures for spam filtering. To fight the spam, there exist many objective measures that can effectively limit the spam email impact on the end users (subjects). However, these objective measures fail in many situations and they can have a level of accuracy. They suffer from the false positive and the true negative problems. Where a non-spam email can be classified as spam email or the filter can classify a spam email as legitimated email (non-spam). The spam problem is very complex one. An ideal anti-spam solution cannot be found as the spammers constantly try to invent new tricks and mechanisms to bypass the objective anti-spam measures so that their spam emails be classified as legitimated emails.

On the other hand, if we can have a group of qualified users (subjects) to monitor and read all the incoming emails, sure they will be the ideal solution for spam filtering. However, for many reasons this solution cannot be applicable because of the privacy, the scalability, the huge volume of emails that is sent by the spammers, etc.

As we highlighted, objective measures alone cannot fight the spam problem. In addition, no pure subjective measures can be implemented effectively to work in real time. Subjective measures can check only the message content (the email body), but in no means they can check other spamming metrics like a forged sender email. Objective measures can accurately and rapidly eliminate in real time up to 70% of spam emails based on known spamming techniques.

Combining both subjective and objective methods can effectively provide better result. If subjective measure is devised in such a way to maintain the privacy of the users, it can cope with almost all the newly devised spammer's tricks that can pass through the objective measures. In addition, they can compensate (correct) the falsepositive objective evaluations.

It is known that the spammers send the same email to thousands of users. In many situations, spammers send the same email every day or week. This led us to think in a way to devise a smart subjective measure so that the spam email can be viewed only by the first few email users. If, for example, four users agree that the email is spam, the system will automatically moves thousands of same-email spam to the users' spam folders. That means that only the first few people will view the spam email and all the rest of the users who received the same email will not have to view it.

However, things are not easy as that. Spammers use special bulk mailing systems and mail merge so that the same email message will have different attributes. For example, they may change the Subject of the email per user. Some spammers include HTML image or Hyperlink links to track the users. Thus, for objective measures they are completely different emails and may seems to be legitimated because they are targeted to specific user.

There is a great challenge to eliminate the customizations the spammers inject into a spam email and to correlate them objectively. When the subjects (users) view the same-email spam copies they will see them as identically. But how to identify and remove the customizations attributes objectively to correlate this same-spam email is really very difficult.

We have devised an approach that is very effective in this matter. To protect the privacy of the user (the email receiver), and to try to correlate all the emails sent from the same spammer to thousands of users, we proceed as follows. Simply remove everything from the email message except the Sender Email address and the filtered message body. In the filtered message body, we keep everything (text) except the hyperlinks (for the images and the links). For these links, we keep only the hostname of the URL. The rest of the URL is removed. Traditionally, spammers tracks the users using the GET parameters which are included in the URL after the "?" (See Examples 4-6 in**Table 1**). However, modern URL rerouting techniques put the parameters in the folder parts of the URL. Example of such techniques are given bellow (Examples 1-3 in **Table 1**). The parameters in these URLs contains information about the target user (usually each email address has a database record in the spammers' servers to know if he/she is active or not, the links he/she is interested in, the location of the user, etc.) In addition, it contains the spam campaign identifier hosted by the spamming infrastructure.

Therefore, in order to be able to correlate the same spam message, we have to remove all the link customization from the body text. This can be done by removing the tracking information in the link. Thus, we keep only the host part of the URL and remove the rest of the URL.

All the other email messages fields can be modified per user by the spammer mailing system. Example of these fields are the Received field (from which mail server at which time), the Receive Date, the Subject, Message-ID, Content Length, To Field, etc.

The proposed technique for spam filtering is as follows:

- An application has to be developed that allows any user to access all the email accounts possessed by the user. This application must have a central database to store the user's information and the email message information.

- Apply the user defined rules and filters to the newly arrived emails. This can include the use of the white, gray, and black lists. It can also combine several rules like "if the subject contains these words and the recipient fields contain some email addresses, and then move the email to this specific folder". The rule or filter action can be to leave the new email in the Inbox, move it to another folder, or simply to delete it.

- On the arrival of any email message, first the traditional spam filtering techniques are applied to filter these emails. No content-based objective filtering is required because they are time consuming and inaccurate. We are analyzing the spam emails since three years. About 70% of the spam emails can be accurately identified using these traditional spam filtering techniques.

- When the user open the Inbox folder of any email account and open any received email that is not filtered using the previous steps, the user can judge the email content whether it is spam or not. In the case when the user judge that the email content is spam, he/she simply clicks on a button labeled “This is a spam”. The application takes this into account and creates (if not already voted by another user) a record for this email. If several users agree that this email content is a spam, then all the other similar emails will be considered as spam. Matching (the meaning of similar) is based on the neutralized body text explained before.

- In this case only the first few users are forced to see the spam emails in their Inbox. For the rest of the users the email is placed in the Spam folder. Moreover, it is known that the spammers periodically send the same spam email (every day, week or month). Thus, the users will not be annoyed by these spam emails that cannot be cached by the objective spam filtering methods.

- On the other hand, if the user opens the Spam folder and finds that any email is not a spam, he/she clicks on a button “This is not a spam”. Then in this case, the application decreases the “Spam weight”. If the “Spam weight” is bellow some value, any similar email is considered as legitimate and placed in the Inbox. If the “Spam weight” is greater than certain value, all the similar incoming emails are considered as spam. If the “Spam weight” is between these thresholds, the email is placed temporally in the “Gray folder”. Periodically, the contents of “Gray folders” have to be reevaluated by the system to decide if the emails are spam or not.

Table 1. Examples of real tracking URL.

Example #	Link (URL)
1	http://ninjaasteroid.com/5264627172690823 72363.UEOVHRA.YTH4D92/197417/141514/3024-006-2-5/b01491b120f69f9534c6ba8ce7106851/euopfedb.5IS6GKTK
2	http://fandragontastic.com/1208108828270473 9366.ABKPNZE.CVNOL3L/591323/140557/3472-006-2-5/3b6e615291cb5839bb2928ac038fcbae/luy56abk.D639K5PO
3	http://click.countrybaby.net/PmgEoCSeXsETlZMEJXUwDzRdWKfrnQEWRijeuRyCDmHRBmlVlrqmdWGihGzV?&n=1585615822&h=a9757f2e2a4885267e6c90e108572ebbf32d26ef
4	http://engine.gtsmobidistributed.com/www/delivery/ck.php?oaparams=2__bannerid=38413__zoneid=1633__cb=d25aa30b4c__oadest=http%3A%2F%2Fwww.mporn.com%2F%3Futm_source%3Dgtsredirects%26utm_medium%3Dcpm%26utm_campaign%3Dmobile_redirects
5	http://click.vegasvapor.net/waqCcEoxTjuEeSbiNLIzRGhJlmKEGRzglvpDRYsnRJCPaEcNMdWGEEWGR?&n=1585581499&h=7f5415cf181e3cf08cbcac6837aa1dc534916258
6	http://www.x3track.com/click.track?CID=209129&AFID=21845&ADID=751949&SID=

To avoid the attacks to the system by the malicious users, a confidence factor has to be associated to each user. If the user makes correct voting (as the average votes of the email), the confidence factor is increased, otherwise it is decreased. If it is bellow some threshold, then the user's voting will not be considered.

4. EMAIL CLIENT APPLICATION

In this Section, we present an Email Client Application (ECA) that we developed that can in addition to the functions offered by the classical email systems offer the proposed anti-spam solutions. We started from the fact that Internet users already have their email addresses (email accounts) and they are using them from long time ago. Any solution forcing the users to have different email address might fail as people do not prefer to change theirs as they are known by their correspondent users. Another fact is that many Internet users have more than one active email accounts (one for the work, one private, and one for web form registration or signing up, etc.). Thus, it is highly preferable for such users to have all their email accounts be read from a single web-based applications so that they can reach all these email accounts and manage them from a centralized location from anywhere in the world at any time without having to install special software.

We have successfully developed a web-based application that combines both the traditional objective measures and a clever subjective measure. Email Client Application (ECA) that we have developed implements the methodology provided in Section 3 and hence can be used by any user to access his email account (s) from a single location from the web browser. It in addition integrates the anti-spam features for all the users' email boxes provided that these accounts have Mail Access Protocol (IMAP) or Post Office Protocol (POP3) access. It includes both the objective (traditional) spam filters as well as a clever subjective spam filtering methodology. This developed Application included the following features.

- Open source software. All the source codes of the application are provided. It can be obtained directly from the system website from the following link (http://aspam.asites.org/packages.zip). This allows any developer to add any feature to the existing one and use it or send us the patch so that we include it in the system.

- To pace the cloud computing era, we developed the system as web application that can be accessed from the Internet using any web browser. It can be accessed from anywhere in the world, at anytime, from virtually any device (Desktop computer, laptop, Smartphone, etc.).

- The main power of the system consists in the use of the traditional objective spam filtering combined with subjective approach. In such a way, the system will be become more and more reliable and effective as the number of users increases. Up to 70% of the spam emails can be filtered using the traditional spam filtering techniques. However, the rest of the cases, especially the content have to be done based on the enduser contributions. Once the end-user opens a new email he can click on a button to say if it is spam or non spam as well as the spam category of the message. This information will help the system to automatically move any quasi-similar email to the user spam folder for the rest of the users.

- The system is developed using the latest MS web development technology such as MS. Net framework v4.0, ASP.NET, VB.NET, etc.

- The developed application can be hosted on any web hosting environment having Internet Information Services (IIS) web server and MS SQL server 2008. Any company or a group of users can obtain the application with its source code, host it on their servers if they prefer to work locally (intranet of a company or an enterprise). We provide more sophisticated solutions (the server edition of the system), where companies/enterprises/groups can add their users to our system and manage their users with all the included security measures. For more details about that see the Email Server Application developed in this project Section 5. We have also developed several web services that can allow all the private installations of our system to interact with the global centralized one. For more details about that see the Web Services developed in this solution in Section 6.

- Any user can easily create an account with a minimal sign-up form on the system. The user can then setup the account by providing his/her existing mailbox account. The user can setup unlimited number of accounts he/she owns. The system does not provide any new email account for the user, but the user will provide the access information for his/her existing email accounts. For example, if the user has an email on yahoo, another on Gmail, another on Gmx, another on his work email server, etc.

- When viewing the list of emails in any folder in any email account, the user has two modes or options: the Lightweight mode and the full processing mode. In lightweight mode, no spam analysis takes place; the headers only are fetched form the remote email servers. No data is stored on the system database. This mode may be fast in the beginning because no email body is downloaded nor spam processing. However, the user can have all the other classical features like moving the emails to another folder or deleting the emails. In addition, in this mode, the user can classify any email he/she opens as spam, gray or non-spam. His rating and spam-classification on a message will be taken into account. In the Full processing mode, the unread emails in the Inboxes are thoroughly examined using the objective and the subjective spam filtering techniques. Email contents (except the attachments) are fetched from the server only once, after being processed, they are stored on the system internal database. Viewing the emails using this mode may by slow at the beginning depending upon the number of non-fetched emails. After fetching the emails, reviewing the email folders will be very fast.

- When the user open any email in any account, he/she can classify it as Spam, Gray, or legitimated one. In the case when classified as spam, he/she can have the ability to categorize it. In the case when it is classified as a spam, lengthy processing operations take place and the email moves to the Spam folder.

- The system supports the Mail User Agent (MUA) functionality, where users could compose there emails and sent it via the Mail Delivery Agent (MDA) via the SMTP protocol. To avoid classifying the emails sent from the global system as spam by the recipient emailing systems, the system will not send directly from the system SMTP server, but using the SMTP authenticated email account of the user (set up in the email account setup process).

- The system supports Mail Retrieval Agent (MRA) functionality, where users could retrieve their emails from the email server(s). MRA supported are both Internet Mail Access Protocol (IMAP) and Post Office Protocol (POP3). In the case when IMAP protocol is used, the user can access all the email folders in that account. He/she can access all the old emails, move the emails between accounts, create/delete email folders, etc. The system supports the access to the built-in folders such as Inbox, Sent emails, Trash, Drafts, Spam. In addition, any user-defined folder is also supported (given that IMAP protocol is used). However, POP3 email accounts can access the Inbox folder only; no folder management can be done. Depending on the user email configuration, the user can access the old email or not.

- The system provides support of all the security and the encryption standards, including the support for Secure Socket Layer (SSL) protocol, Transport Layer Security (TLS), S/Mime, APOP, NTLM/GSSAPI, and FIPS etc. When setting up an email account, using the port number, the system will automatically indentify the security and encryption standard that will be used.

- The system supports all the email servers providing IMAP and POP3 access, where any user who has a mailbox on any server supporting IMAP or POP3, could read that account from the system via IMAP or POP3. In addition, the user could receive all the emails on the system and could also send the email from the system. Examples of such email servers are: MS Exchange, Lotus Notes, GroupWise, IIS SMTP/ POP3, IMail, MailEnable, AxiGen, SmarterMail, SurgeMail, MDaemon, Kerio, CommuniGate, hMailServer, Exim, Postfix, Sendmail, Qmail, Courier, Dovecot, Cyrus, Zimbra, Gmail, Hotmail, Yahoo... and any POP3/SMTP/IMAP compliant servers.

- A wide range of filters and actions is available. Each filter can have several rules: Sender email, the recipient (To and CC), the Subject. For each of these fields, the user can select: Start with, contains, end with, or exactly equal. If all the rules apply to any newly arrived email address, there are three actions: Leave it in its folder, delete it or move it to another folder in the same account or to a different email account for the user.

- Any user account has three built-in lists: Black list, Gray list and White list (per user lists). The user can create any other lists (per user lists). There is a whole list management configuration page on the system. User can create, rename and delete any of the userdefined lists. However, he/she cannot modify any of the built-in lists. Each list has an associated folder. On the creation of the list, the user must select the associated folder for it (the user has to select the Email account first, and then he/she has to select the folder from all the available folders under this email account). The user then can have full control over the content of each list. The entries of the list can be a complete email address, a domain address, or any part of the email. On the arrival of any new email, when the system process it, if the sender email address matches any of entry of the list, then that email will be moved directly to the associated folder in the associated email account for that user.

- On the arrival of new email messages, the order of processing will be as follows:

a) First if the sender email/sender domain/or part of the sender email exists in any of the user lists, then the email message will be moved to the associated folder in the associated email account of the user.

b) Otherwise, the filter rules are checked. If all the rules are satisfied for a specific filter, then the associated filter action will be executed. Actions may be to move the email to a specific folder in a specific email account or to delete the email completely, to move it to the gray folder or the spam folder of the user.

c) Otherwise, the email message is checked if it has been reported by any user before as being spam or not. If so, check the spamwieght of the previously reported email, if it is greater than the SpamThreshold then move it to the Spam Folder of the user, otherwise move it to the Gray Folder of the user.

d) Otherwise, apply the objective spam filtering rules (valid DNS, rDNS, DKIM, etc.). If the email pass all the tests, then the email is left in its Inbox folder in that email account, if on the other side one of the tests fails, then the email is moved to the Spam Folder of the user.

e) Regularly, emails that have been moved to the gray folder automatically are reevaluated, if the Spamweight is more than the threshold, then move it to the Spam folder, otherwise move it to the inbox of the user. This is to avoid stalling an email long time in the gray folder and it could be important for the user and not important to the first few users who reported it as spam. If the user find that it is spam, he can report that and the Spamweight will increase. This can improve the accuracy and eliminate the need of other users to view a spam email.

- Security is of great concerns. Thus the developed application is secured and proof-tested against the known security attacks such as, SQL injection attack, cross sight scripting, etc.

5. COOPERATIVE ANTI-SPAM EMAIL SERVER APPLICATION

The second part of the developed system is Cooperative Anti-Spam Email Server Application (CASESA). It is complementary to the Email Client Application (ECA), refer to Section 3. The server can automatically classify any email for any email server that integrates our web services for any incoming email. This is based on the subjective evaluations of the emails carried out by the users of the developed ECA, or by using any email client application with the help of the Web Services that have been developed (See Section 6).

Using the innovative technology named as the web service (part of the cloud computing), we have developed a web service that allows any existing or new email server or email client application to ask for classifying an incoming email message. In addition, for companies or user groups and lists, the system administrator can simply create accounts for the users in the system in one action. After that, all the users can benefit from the services and features offered on the system. In addition, they can participate in the subjective spam evaluation process to enrich the system and to improve the accuracy of the subjective rating.

As said before, it is known that all the existing objective anti-spam filters are not accurate and thus give false positives and false negative classification. The most accurate one is the subjective method. By allowing the user to give his feedback about the received email and by building a huge server application hosting all the accounts of the users, then once a new email is received by any user and opened, if the user subjectively classify it as a spam by clicking on the spam button, then the client application will send this email message to server saying that "this is a Spam".

The server then take this into account and any new message will be checked against the existing database of the subjectively-classified email messages. If it is found, then the server will place the email in the Spam folder of the user. Usually, the spammers send the same email to many users (bulk) but changing only the destination. Thus, in this scenario, only the first user will open the spam email and any other user in the system will not be annoyed by that email with high confidence as the classification is done by human. Our developed system included the following features.

- Opened source software, where all the source codes are available on the project website.

- The developed application can be hosted on any web hosting environment having Internet Information Services (IIS) web server and MS SQL server 2008. Any company or a group of users can obtain the application with its source code, host it on their servers if they prefer to work locally (intranet of a company or an enterprise).

- The system is developed using the latest MS web development technology such as MS. Net framework v4.0, ASP.NET, VB.NET, etc. In addition, we have used the standard Web services to communicate with the system as well be explained in Section 5.

- The system is intended mainly for servicing the email servers to help them automatically obtain objective and subjective email filtering for the users having email accounts on these servers and/or allowing a single person to create and manage all the existing users on these servers. For this reason, a single user must be nominated by the enterprise, the company, the campus, or any establishment to manage the users (adding them to the system, validating them, disabling any one of these users, etc.). We refer to this user as the ServerUser. ServerUser must contact the system Superadmins to create a server manager account for them, upon approval, an account will be created for the ServerUser and an email will be sent to his account to activate his account. We have chosen this methodology to avoid the Spamming attack to the system and to allow the service for the serious people only.

- Once the ServerUser account is created and activated, he/she can log into the system and create client access accounts on the system for all the users on the email serves in that enterprise, the company, the campus, or any establishment. The simplest way for doing that is by composing all the user information in a single Excel file and importing them from the system. We have prepared a template excel file that can be used for that purpose. One very important thing to be noted is that, the ServerUser must enter the Email Server access information (the SMTP host name, the SMTP port number, the email access server type, the email access host name, and the email access port number). This information plus the specific user information imported from the Excel file will be used for creating email client account for every user on the system.

- Upon the creation of user accounts of the users in previous step by the ServerUser, anyone of these users can log into the system (The Email Client Application described in details in Section 3).

- All the features provided in the Email Client Application are integrated on the developed Cooperative AntiSpam Email Server Application. There is no need to repeat them here, for more details, please see Section 3.

- The system developers can make use of the developed Web Services (cf. Section 6) allow these users to remotely provide their subjective voting on the emails they receive. In addition, the system developers can include these web services to allow their email servers to automatically classify any incoming email for these users as spam or non-spam.

- Each user is assigned a confidence factor to avoid injecting bogus data by malicious users. If he/she reported a spam email with other users, this factor will be increased, otherwise, it is decreased. If the confidence factor of the user was less than a threshold, then this user is placed on the black list and all his/her future judgments will be ignored.

- The system periodically calculates and updates the Confidence Factor for all the Email Users on the system. As this step is a lengthy process and usually the usual user can vote on around 10 emails daily, there is no need to shorten the period. We have chosen that this process can be done once per day (every 24 hours). In addition, for the optimization reasons, we have chosen to make the processes be executed by the database server as a stored procedure and called by the web server once per day. Again for optimization reasons, only the users who voted on some emails during the current period are processed (their votes are stored on a table during that period), then this table is flushed. The calculation of the confidence factor of the user is done in the database table as a Computed Column based the # of Correct Votes and the # of Wrong Votes of the user using the following formula. The constant 0.000000001 is to avoid the divide by zero problem and to avoid using an additional test condition. If the user provides less than 30% correct votes, then his confidence factor is zero (as if he/she is panned or black listed). Thus all his/her next votes will not be considered. There is no need to manually remove the user from the black list. The user can help him/her self by improving his subjective evaluation. As the script is executed daily, once he/she gives better votes (more than the 30%

threshold), his/her opinion on filtering email messages subjectively will be considered weighed to the value of the confidence factor.

```
(case when [NoCorrectVotes]/
(([NoCorrectVotes]+[NoWrongVotes])+
(0.000000001))<(0.3) AND
([NoCorrectVotes]+[NoWrongVotes])>(100)
then (0) else [NoCorrectVotes]/(([NoCorrectVotes]+
[NoWrongVotes])+(0.000000001)) end)
```

- Once an email arrives on any of the Inbox of any of the Email users, if the user opens it and classify it as Spam, in this case, it will be placed on the Spam folder of that user. However, we cannot be sure that this email is Spam (we cannot base our judgment on a single user). As a result, when any user opens his Inbox folder in any email account, the new emails will be processed as follows. If the email is previously rated by any user as Spam, we check the SpamWeight. If it is >4, then the email is moved to the Spam folder, else it is moved to the Gray folder. If not previously rated by any user, the email remains in its folder (after executing the user defined filtering rules). The SpamWeight for each message is calculated based on the Confidence Factor of the voting user. If the user rate the email as Spam, then the SpamWeight is incremented by the Confidence Factor of the voting user, else, the SpamWeight is decreased by the Confidence Factor of the voting user. One problem is that once the user opens the any of the Email folders, the email will be classified based on the previously mentioned rule and it will remain in that state even if its SpamWeight is updated with time by other users. It could remain in the Inbox even if it is Spam and it could remain in the Gray or the Spam folder even if it is legitimate email. To solve this problem, we have designed a stored procedure that executes every 10 minutes on the database server that periodically updates the states of the emails of the users based on the overall user voting's. To avoid updating the status of the emails that have been moved by the user manually, we apply this procedure to those emails that are not moved manually by the user using the isMovedByUser flag.

6. ANTISPAM WEB SERVICES

We have developed a complete solution consisting in two main components: the first one is the complete Email Client Application (refer to Section 3 for more details) and the second one is the Cooperative Anti-spam Server Application (refer to Section 5 for more details). We know, however, that there are many people who are not willing to migrate to our Email Client Application, even if it has many features. For example, Gmail, yahoo, live, hotmail, enterprise email servers like Exchange Servers, Exim, etc email users are billions and cannot migrate to ECA. Even if the system support the access to these email servers email accounts and their email boxes can be accessed from that system, the end users and even the owners of these email solutions may not prefer to migrate. From another point of view, the subjective anti-spam filtering accuracy depends on the number of the users voting on the viewed email. Even if we have used the confidence factor to eliminate or minimize the unreliable and the malicious users, the more the email voters, the more accurate the results will be.

Because of all these reasons, we have developed some standard web services that can be used by almost any email access client software and almost any email server software. To illustrate the idea, let's assume that a company would like to make use of the spam filtering solutions provided by the developed solution which is installed on the (http://www.aspam.org). In this case, a developer must add a simple code to the email server such that on the arrival of any email, a request to our system will be sent to know if that email is previously received by any users and classified as spam or not. The response from our system to the company email server would contain much other useful information.

The company email server then based on the received response can decide to place the email in the Inbox of the user or to move it to the Spam folder. This action is completely transparent to the end user. Taking this into account, almost all the emails servers can integrate this functionality to their servers to provide the subjective and objective email filtering solution for their end users. This is exactly the first web service that we have developed. It is named as ClassifyNeutralEmail. We know that the privacy is of great importance for the users. Additionally, not all the email information is required by the system to be able to match the email in the database of the system. After several studies, we concluded that only the email sender and the message body are needed (as described in Section 5). Other information such as the recipients (To and CC fields), the subject, the receive time, the sent time, MessageID, etc. can be variable for the same spam email from user to user. In addition, the spammers can change these fields per users. Thus to increase the accuracy and to maintain the privacy of the users, no need to send these values and no data will be stored on the system on the recipient of the request from the servers. Regarding to the Body contents, it has to be neutralized (as described in Section 3).

On the other hand, what about the existing email clients applications such as Ms outlook, Thunderbird, Webbased email access clients (e.g. Gmail, yahoo, Outlook Web Access, etc), and the Smartphone email access programs (Android, Iphones, IPad, Nokia, Windows Mobile, etc)? Billions of users are using these email access programs and they liked it and cannot change to another solution easily. We can hit two birds by a single stone using the following procedure. We have developed several standard web services that can be directly integrated with almost any existing email access clients (Mail User Agents) to allow the billions of these end users to participate into the subjective email filtering problem.

The developers can simply add two simple buttons on this Mail User Agents (MUA) if they do not exist before. The first one will be shown when the user opens and email found in the Inbox labeled "This is a Spam message". When the user open an email and decide that this is a spam and he/she clicks on that button, a call to reportSpamEmail web service will be executed. The message will be executed on the remote server (Our system). An intelligent search in the database will be performed to find if the message is reported previously by any user. If it is found, then update the message record by incrementing the number ReportedAsSpam counter and increase the SpamWeight by the Confidence Factor of that user. If the email is not found in the database, then a record for it is created and the voting counters for it are initialized.

Similarly, when the user is viewing the contents of his/her Spam folder on the MUA, he/she can decide that an email is not a spam. Thus, a developer can update this MUA software by adding a button labeled "This is not a spam" as an option. When the user clicks on that button a call to the web service named reportNonSpamEmail will be executed. The message will be executed on the remote server (Our system). An intelligent search in the database will be performed to find if the message is reported previously by any user. If it is found, then update the message record by decrementing the number ReportedAsSpam counter and decreasing the SpamWeight by the Confidence Factor of that user. If the email is not found in the database, then a record for it is created and the voting counters for it are initialized.

For security reasons, all the implemented web services cannot be used anonymously. Reporting a spam message or reporting non-spam email message must be done using a previously registered user. This is to avoid the attacks from the spammers or the malicious users. In addition, this is a must for calculating and updating the confidence factor of the reporting user. It may be thought that forcing all the users to create an account on the system can be difficult task, but using the Cooperative Anti-spam Server Application (refer to Section

5, Page 17 for more details), all the users accounts can be created in a single operation, by importing their details from a single excel file.

For the performance and availability issues, and to be able to service millions of users and thousands of email servers and email access clients, the system must be installed on a server farm.

Another factor is the intelligent technique in the search in the database. The database tables could contain millions of records. Thus, for every incoming email a search must be done on the entire database. Even using the Indexing technique cannot help because that can make inserting and updating very slow. We are using a method which consists in using Hashing technique combining both the Sender + the body. We search based on the hash code (MD5). The hash code can be indexed and as it is constant size, it cannot make performance issue on the Insert and the Update. We know that several different emails can have the same hash code, thus, we iteratively try to scan all the messages having the same hash code to find the exact match.

7. USING THE ASPAM WEB SERVICES

We briefly explain here how to integrate these web services within the email applications: Mail User Agent (MUA) and Mail Servers. Details on how to integrate the web services in any application is outside the scope of this paper. It is extremely simple for a professional developer. There exist many resources on the Internet showing how to use them. These web services are provided with the source code of the developed applications. Several sample applications are developed and can be obtained from the source code of the whole system.

These web services are hosted in the web site of the project under the following link: http://www.asites.org/aspamServices.svc. If you navigate to that link, you will simple see a very simple how-to use these services. The source code of the web services are also provided and can be used in a stand-alone installation within the whole system. More details and description of the web services and the sample application code description are provided in an extended technical report [35].

8. CONCLUSIONS AND FUTURE DIRECTIONS

In this paper, a new technique for spam filtering that combines both the traditional spam filtering methods and a smart cooperative subjective spam filtering method is presented. A developed email client application (ECA) that implements the proposed technique is described. This application provides the traditional email services in addition to the spam filtering features. It makes use of the user voting on the incoming emails to classify the rest of the similar emails for the other users automatically. The second developed application that is described in the paper is the Cooperative Anti-Spam Email Server Application. It complements the ECA and provides several features for the enterprises and the email service providers. Finally, a set of standard web services that we have developed are presented. One of them allows any existing email server to request for filtering any arrived email as spam or not based on the subjective spam filtering technique provided here. The other web services allow the users of the existing email client applications to participate on the subjective spam filtering activity on the centralized system. The system is validated and tested. The accuracy of the system depends on the number of users using it. As more users use the system, its accuracy increases

REFERENCES

1. W.-F. Hsiao and T.-M. Chang, “An Incremental Cluster- Based Approach to SPAM Filtering,” Expert Systems with Applications Journal, Vol. 34, No. 3, 1975, pp. 1-28.

2. N. Leavitt, “Vendors Fight Spam’s Sudden Rise,” IEEE Computer Magazine, Vol. 40, No. 3, 2007, pp. 16-19.

3. L. M. Spracklin and L. V. Saxton, “Filtering Spam Using Kolmogorov Complexity Estimates,” The Proceedings of the 21st International Conference on Advanced Information Networking and Applications Workshops, Niagara Falls, 21-23 May 2007, Vol. 1, pp. 321-328.

4. Y.-M. Wang and M. Ma, “Strider Search Ranger: Towards an Autonomic Anti-Spam Search Engine,” Fourth International Conference on Autonomic Computing, Jacksonville, 11-15 June 2007, pp. 32-32.

5. M. N. Marsono, M. Watheq El-Kharashi and F. Gebali, “Binary LNS-Based Naive Bayes Inference Engine for Spam Control: Noise Analysis and FPGA Implementation,” IET Computers & Digital Techniques, Vol. 2, No. 1, 2008, pp. 56-62.

6. K. Saraubon and B. Limthanmaphon, “Fast Effective Botnet Spam Detection,” Fourth International Conference on Computer Sciences and Convergence Information Technology, Seoul, 24-26 November 2009, pp. 1066-1070.

7. Z. H. Duan, P. Chen, F. Sanchez, Y. F. Dong, M. Stephenson and J. Barker, “Detecting Spam Zombies by Monitoring Outgoing Messages,” IEEE INFOCOM 2009, 19-25 April 2009, pp. 1764-1772.

8. Z. H. Duan, P. Chen, F. Sanchez, Y. F. Dong, M. Stephenson and J. M. Barker, “Detecting Spam Zombies by Monitoring Outgoing

Messages," IEEE Transactions on Dependable and Secure Computing, Vol. 9, No. 2, 2012, pp. 198-210.

9. E. Levy, "The Making of a Spam Zombie Army: Dissecting the Sobig Worms," IEEE Security & Privacy Magazine, Vol. 1, No. 4, 2003, pp. 58-59.

10. Z. Li and H. Y. Shen, "SOAP: A Social Network Aided Personalized and Effective Spam Filter to Clean Your EMail Box," Proceedings of IEEE Infocom, Shanghai, 10- 15 April 2011, pp. 1835-1843.

11. J. A. Zdziarski, "Ending Spam—Bayesian Content Filtering and the Art of Statistical Language Classification," 5th Edition, No Starch Press, San Francisco, 2005.

12. I. Androutsopoulos, G. Paliouras, V. Karkaletsis, G. Sakkis, C. D. Spyropoulos and P. Stamatopoulos, "Learning to Filter Spam E-Mail: A Comparison of a Naive Bayesian and a Memory-Based Approach," Proceedings of the Workshop: Machine Learning and Textual Information Access, 2000, pp. 1-13.

13. M. Sahami, S. Dumais, D. Heckerman and E. Horvitz, "A Bayesian Approach to Filtering Junk Email," Learning for Text Categorization—Papers from the AAAI Workshop, 1998, pp. 55-62.

14. D. Karthika Renuka, T. Hamsapriya, M. Raja Chakkaravarthi and P. Lakshmi Surya, "Spam Classification Based on Supervised Learning Using Machine Learning Techniques," International Conference on Process Automation, Control and Computing, Coimbatore, 20-22 July 2011, pp. 1-7.

15. H. Drucker, D. Wu and V. N. Vapnik, "Support Vector Machines for Spam Categorization," IEEE Transactions on Neural Networks, Vol. 10, No. 5, 1999, pp. 1048-1054.

16. C.-Y. Tseng and M.-S. Chen, "Incremental SVM Model for Spam Detection on Dynamic Email Social Networks," International Conference on Computational Science and Engineering, Vancouver, 29-31 August 2009, pp. 128- 135.

17. S. Suwa, N. Yamai, K. Okayama and M. Nakamura, "DNS Resource Record Analysis of URLs in E-Mail Messages for Improving Spam Filtering," IEEE/IPSJ 11th International Symposium on Applications and the Internet (SAINT), Munich, 18-21 July 2011, pp. 439-444.

18. A. Khanal, B. S. Motlagh and T. Kocak, "Improving the Efficiency of Spam Filtering through Cache Architecture," 15th International Symposium on Modeling, Analysis, and Simulation of Computer and Telecommunication Systems, Istanbul, 24-26 October 2007, pp. 303-309.

19. T. R. Surmacz, "Reliability of E-Mail Delivery in the Era of Spam," 2nd International Conference on Dependability of Computer Systems, Szklarska, 14-16 June 2007, pp. 198-204.

20. B. Hoanca, "How Good Are Our Weapons in the Spam Wars?" IEEE Technology and Society Magazine, Vol. 25, No. 1, 2006, pp. 22-30.

21. E. Rabinovitch, "Readers' Comments on SPAM," IEEE Communications Magazine, Vol. 40, No. 11, 2002, pp. 20-24.

22. J. Yan and P. L. Cho, "Enhancing Collaborative Spam Detection with Bloom Filters," Proceedings of the 22nd Annual Computer Security Applications Conference, Miami Beach, December 2006, pp. 414-428.

23. "Razor: A Distributed, Collaborative, Spam Detection and Filtering Network," http://razor.sourceforge.net

24. S. A. Elsagheer Mohamed, "A Solution for Fighting Spammer's Resources and Minimizing the Impact of Spam," International Journal of Communications, Network and System Sciences, Vol. 5, No. 7, 2012, pp. 416-422.

25. X. Carreras and L. Márquez, "Boosting Trees for Antispam Email Filtering," Proceedings of 4th International Conference on Recent Advances in Natural Language Processing, 2001, pp. 58-64.

26. M. R. Islam, W. L. Zhou and M. U. Choudhury, "Dynamic Feature Selection for Spam Filtering Using Support Vector Machine," 6th IEEE/ACIS International Conference on Computer and Information Science, Melbourne, 11-13 July 2007, pp. 757-762.

27. W. W. Cohen, "Learning Rules That Classify E-Mail," Proceedings of AAAI Spring Symposium on Machine Learning in Information Access, Stanford, 25-27 March 1996, pp. 18-25.

28. X.-L. Pang, Y.-Q. Feng and W. Jiang, "The Compensation Strategy of Unseen Feature Words in Naïve Bayes Text Classification," Journal of Harbin Institute of Technology, 2007.

29. A. Taweesiriwate, B. Manaskasemsak and A. Rungsawang, "Web Spam Detection Using Link-Based Ant Colony Optimization," IEEE 26th International Conference on Advanced Information Networking and Applications (AINA), Fukuoka-shi, 26-29 March 2012, pp. 868-873.

30. S. Mohamed, W. Ata and N. Darwish, "A New Technique for Automatic Text Categorization for Arabic Documents," Proceedings of 5th International Conference on Internet and Information Technology in Modern Organizations, Cario, December 2005.

31. H. Q. Zuo, X. Li, O. Wu, W. M. Hu and G. Luo, "Image Spam Filtering Using Fourier-Mellin Invariant Features," IEEE

International Conference on Acoustics, Speech and Signal Processing, Taipei, 19-24 April 2009, pp. 849- 852.

32. J.-H. Hsia and M.-S. Chen, "Language-Model-Based Detection Cascade for Efficient Classification of ImageBased Spam E-Mail," IEEE International Conference on Multimedia and Expo, New York, 28 June-3 July 2009, pp. 1182-1185.

33. B. Biggio, G. Fumera, I. Pillai and F. Roli, "Image Spam Filtering Using Visual Information," 14th International Conference on Image Analysis and Processing, Modena, 10-14 September 2007, pp. 105-110.

34. H. B. Aradhye, G. K. Myers and J. A. Herson, "Image Analysis for Categorization of Image-based Spam EMail," Proceedings of the 8th International Conference on Document Analysis and Recognition, Korea, 29 August-1 September 2005, pp. 914-918.

35. S. A. E. Mohamed and M. M. Tezeghdanti, "Spam Detection and Categorization for Electronic Arabic Messages: Project Final Technical Report," Technical Report, 2012. http://aspam.asites.org/0Files/TechnicalReport-FinalV3.pdf

CHAPTER 4

Personalized Tag Recommendation Based on Transfer Matrix and Collaborative Filtering

Shaowu Zhang[1,2], Yanyan Ge[1]

[1] School of Computer Science and Technology, Dalian University of Technology, Dalian, China

[2] School of Computer Science and Engineering, Xinjiang University of Finance and Economics, Urumqi, China

ABSTRACT

In social tagging systems, users are allowed to label resources with tags, and thus the system builds a personalized tag vocabulary for every user based on their distinct preferences. In order to make the best of the personalized characteristic of users' tagging behavior, firstly the transfer matrix is used in this paper, and the tag distributions of query resources are mapped to users' query before the recommendation. Meanwhile, we find that only considering the user's preference model, the method cannot recommend new tags for users. So we utilize the thought of collaborative filtering, and produce the recommend tags based on the query user and his/her nearest neighbors' preference models. The experiments conducted on the Delicious corpus show that our method combining transfer matrix with collaborative filtering produces better recommendation results.

KEYWORDS

Tag Recommendation, Collaborative Filtering, Transfer Tensor

1. INTRODUCTION

With the rapid growth of the web and social networks, more and more people like to share their information on the Internet, including pictures, videos and even political opinions. This leads to the flourish of the web data, and increases difficulties for people to organize and search resources they want. In order to address this challenge, social tagging has attracted more attentions. Delicious, Bibsonomy, Flickr, and Last.fm are all the popular social tagging online services recently.

Tagging allows users to label content by assigning freely chosen keywords (tags). In such open environment, tagging has proven a superior alternative to traditional categorization techniques due to its flexibility that enable users to choose labels that match their real tastes. And the increase of the descriptive keywords on resources is convenient for people to search, especially in the case of multimedia. Different online tagging service distinguishes in the resource type, but they are similar in essence, such as Delicious supports the tagging on webpages, Bibsonomy allows to tag webpages and journal papers, and people can tag pictures and the music in Flickr and Last.fm, respectively [1]. This kind of community services that centers on the tagging of resources is also called folksonomies. These communities have become a valuable source of information, since they bundle the interests, preferences of thousands or millions of users.

Tag recommenders support a user during the posting process by suggesting potentially relevant tags. Because of the stability of a user's tag vocabulary, it is effective to predict the future tagging behavior of a user according to the past. Previous works have showed that the description of the same object is different due to users' customs and knowledge. So in the social tagging system, each user can generate a set of tags with their own preferences, which is called personalized tag vocabulary. At the same time, tags on the resources are based on the whole tagging system and called folksonomies. The tagging system builds preference models of users and tag models of resources by analyzing users' tagging history, and generates the correlations among tags, users and resources; then recommends the top n relevant tags to users. Depending on whether consider users' preference models or not, tag recommendation comes in two forms, personalized and non-personalized. For a resource, personalized tag recommender systems recommend different tags for different users in contrast to non-personalized tag recommenders. And for a good tag recommender system, personalized recommend should be given preference.

Wetzker et al. [2] introduce a novel tag model that allows deriving mappings between personal tag vocabularies and the corresponding folksonomies. Using these mappings, it can infer the meaning of user-assigned tags and can predict choices of tags a user may want to assign to new items. So this kind of transfer based method can meet the personalized demand of users. But it only considers the preference model and the used tags of a user, and cannot recommend new tags.

In order to solve this limitation, based on the transfer tags, we integrate the collaborative filtering technology and generate a new algorithm. First, we build the personalized transfer matrix for each user; the element in the matrix represents transfer values between personalized tags and folksonomies. Then, we determine the nearest neighbors of a user by computing their similarities. Finally, we recommend tags by considering the user and his/her nearest neighbors' preference models and their similarities together. In addition, for similarities between users, we propose a new method based on the personalized transfer matrix.

The remainder of this paper is structured as follows. We begin with a discussion of related work in Section 2. Section 3 describes our method in detail. And then analyze the experiments in Section 4. Section 5 concludes the work of the paper.

2. RELATED WORK

2.1. Tag Recommendation

Recently, the social tagging system has attracted a lot of attention, and the recommender system based on the social tagging also obtains much more focus.

There are many classical recommendation algorithms. Collaborative filtering based method is traditional and will be described in the next subsection. Resource content based method focuses on the texts or webpages and mainly depends on the key words extraction technology to select the tags. Zhang et al. [3] discussed that different resources and users should adopt different recommend methods, and for those new resources, resource content based method is a choice. Guan et al. [4] took the tag recommendation as a "questions and sort" problem, and proposed a graph-based ranking technology.

Symeonidis et al. [5] proposed to decompose the full folksonomy tensor and estimate missing values using Higher Order SVD (HOSVD). The authors claim reasonably good results on the task of tag recommendation. Rendle et al. [6] [7] reported some researches on applying tensor factorization methods on the tag recommendation.

Folk Rank algorithm is inspired by the seminal Page Rank algorithm [2] [8] -[10] . The basic notion is that a resource which is tagged with important tags by important users becomes important itself. Folk Rank is based on the random surfing model, considering the weight and weight-spreading of edges. The algorithm cannot be applied directly on folksonomies; it needs to be transferred into an undirected tripartite graph including users, resources, and tags.

And also there are many recommender systems consist of different methods. Zhang et al. [2] combined the content based and collaborative filtering algorithm together, the former applied to new resources, the later used to other situations. Zowl et al. [11] designed four tag recommendation strategies by combining the tag co-occurrence, tag aggregation (including vote and sum strategies), and optimized re-ranking method.

2.2. Collaborative Filtering

Collaborative filtering is a widely used and effective recommendation technology. It gets favored by many commercial websites, such as Amazon.com. The underlying assumption of the approach is that users would have similar behavior in the future if they performed similar in the past. In the recommender systems, collaborative filtering comes in two methods, user-based and item-based. The user-based algorithm recommends a user what his/her nearest neighbors interest in. The item-based algorithm recommends a user an item by judging whether the user interests the nearest neighbors of the item.

There are many researches about applying collaborative filtering in the recommender system [12] . Previous works have summarized the approach and proposed to model social tagging systems as tripartite hypergraphs [13]-[15] . Marinho et al. [16] designed a tag recommender system with collaborative filtering method, and results show that it performs better than the other two popular methods. Xu et al. [17] proposed an iterative algorithm based on the collaborative filtering to compute the quality of tags. Koren [18] also combines the collaborative filtering and temporal dynamics in the recommendation.

Gueye et al. [19] [20]. develop a network based parameter-free system to optimize the tag recommendation. Ifada et al. [21] introduce learning to rank methods from information retrieval field to optimize the DCG measure in tag-based item recommendation.

3. TRANSFER MATRIXES BASED PERSONALIZED TAG RECOMMENDATION MODEL

In the social tagging system, different tags are called co-occurrence if they appear in one resource. The paper first makes use of this co-occurrence among tags to build mappings between personalized user tags and folksonomies on resources, and identify the nearest neighbor set N of the user by integrating collaborative filtering technology. Then in the tag recommendation stage, we can map the tag distribution of the query resource to the distributions of tags that the query user and his/her neighbors have used.

The arrangement of the part is as follows. Subsection 3.1 gives the description of the problem. We introduce the build of the personalized transfer matrix and the identification of the nearest neighbors in subsection 3.2. Subsection 3.3 describes the method to recommend tags based on the user and nearest neighbors.

3.1. Problem Description

A social tagging system includes users, resources, and tags, and the tagging behavior of a user reflects the relationship among them. A folksonomy is a tuple F: = (U, T, R, A) where U, T, and R are finite sets, whose elements are called users, tags, and resources, respectively, and A is a ternary relation among them, i.e. $A \subseteq U \times T \times R$, whose elements are called tag assignments (TAS for short). A user and a resource tagged himself/herself together are called a bookmark which can be represented as $B := \{(u,r) \exists t : (u,t,r) \in A\}$.

For the given user and the resource (u, r), we need to compute correlations between a tag and u and r. The higher correlation represents the higher possibility that the user u assigns this tag to the resource r. Then we can recommend the highest tags to the user.

3.2. Build Personalized Transfer Matrix

The relations between resources and tags are represented as a $|R| \times |T|$ matrix X as follows:

$$\boldsymbol{X} = \left[\boldsymbol{x}_1, \boldsymbol{x}_2, \cdots, \boldsymbol{x}_{|R|}\right]^{\mathrm{T}} \tag{1}$$

$$\boldsymbol{x}_r = \left(x_{r1}, x_{r2}, \cdots, x_{r|T|}\right), \quad r := 1, \cdots, |R| \tag{2}$$

where x_{rt} is times that the tag t assigned to the resource r, and $x_{rt} = |\{u | (u,t,r) \in A\}|$. A row vector $\boldsymbol{x}_r$ is a tag distribution of the resource r, and a column vector is a resource distribution of the tag t. After the normalization on rows of the matrix X, we get X':

$$\boldsymbol{X}' = \left[\boldsymbol{x}'_1, \boldsymbol{x}'_2, \cdots, \boldsymbol{x}'_{|R|}\right]^{\mathrm{T}} \tag{3}$$

$$\boldsymbol{x}'_r = \left(x'_{r1}, x'_{r2}, \cdots, x'_{r|T|}\right), \quad r := 1, \cdots, |R| \tag{4}$$

$$x'_{rt} = \frac{1}{\sum_{i=1,\cdots,|T|} x_{ri}} x_{rt} \tag{5}$$

The relation between users' personalized tags and folksonomies is built through resources. If any two tags co-occurrence on many resources, they will have a higher relation. So we set up a mapping between users' personalized tags and folksonomies based on the co-occurrence. For a user u, $\hat{T}$ and T represent the used personalized tag vocabulary of u and the folksonomy, respectively. We then can represent the mapping as a $|\hat{T}|\times|T|$ personalized transfer matrix Y_u, and Table 1 is an example.

$$\boldsymbol{Y}_u = \left[\boldsymbol{y}_{u1}, \boldsymbol{y}_{u2}, \cdots, \boldsymbol{y}_{u|R|}\right]^{\mathrm{T}} \tag{6}$$

$$\boldsymbol{y}_{u\hat{t}} = \left(y_{u\hat{t}1}, y_{u\hat{t}2}, \cdots, y_{u\hat{t}|T|}\right), \quad \hat{t} := 1, \cdots, |\hat{T}| \tag{7}$$

where element $y_{u\hat{t}t}$ is the mapping value between a personalized tag $\hat{t}$ and a folksonomy t. It can be represented as Equation (8).

$$y_{u\hat{t}t} = \sum_{r\in R} \delta(u, r, \hat{t}) x'_{rt} \tag{8}$$

If $(u,r,t)\in A$, it indicates the user u assigned the tag $\hat{t}$ to the resource r, then $\delta(u,r,\hat{t})$ equals to 1.0, or to 0.0. Then

$$y_{u\hat{t}t} = \sum_{r\in R(u,\hat{t})} x'_{rt} \tag{9}$$

where $R(u,\hat{t})$ represents the resources assigned with $\hat{t}$ by the user u.

3.3. Identify Nearest Neighbors

It needs to identify the nearest neighbor set after building the personalized transfer matrix. Here we use the cosine to measure the similarity, and choose the top K as the nearest neighbors of the query user. According to Marinho [16] , we compute similarities under two situations. One is based on the resources with tags assigned by users. Assuming the number of resources is M in total, a user can be represented as an M vector $(p_1, p_2, \cdots, p_M)$, and p_i equals to 1.0, if the user assigned tags to the resource r_i, otherwise is 0.0. The other is based on the tag space model. Assuming there are N tags, then a user is a N vector $(q_1, q_2, \cdots, q_N)$, and q_i is the ratio that the number of the user u use the tag t_i in to the all tags used by the user, which reflects the probability that the user u use the tag t_i.

In addition, we also propose a method to compute the similarity between users based on the personalized transfer matrix. If any two transfer matrixes of users have a high similarity, we can believe that the two users are similar. If both the personalized tag vocabularies of user u and v include the tag t_i, we can define the user cosine similarity based on the tag t_i as follows:

$$sim(u,v)\big|_{\hat{t}} = \cos\left(Y_{u\hat{t}}, Y_{v\hat{t}}\right) \tag{10}$$

$$sim(u,v) = \exp\left(\frac{1}{\beta n}\right)\frac{\sum w(u,t)\, sim(u,v)\big|_t}{\sum w(u,t)} \tag{11}$$

where $Y_{u\hat{t}}$ and $Y_{v\hat{t}}$ in Equation (9) are the transfer vectors of user u and v based on the tag $\hat{t}$. In Equation (11), w(u, t) is the probability that the user u use the tag t as we defined in the subsection 3.2. N is the number of tags that both user u and v have, and the higher the N, the higher the similarity between users. β is a tunable parameter, and we get our best results when it sets to 3.

Table 1. The personalized transfer matrix of a user.

User Tags \ Resource Tags	t_1	……	t_k	……	$t_{\|T\|}$
t_{u1}	0.242		0.135		0.335
……					
t_{uk}	0.521		0.362		0.01
……					
$t_{u\|T\|}$	0.03		0.321		0.201

3.4. Tag Recommendation

For the given query user and the resource (u, r), we combine the thought of collaborative filtering with personalized transfer matrix when recommend tags. This can ensure both the personalization and the generation of new tags. For the query user u, we first get his/her nearest neighbor set N(u) by Equation (11), then mapping the tag distribution of the resource r to the tag vocabulary of user u and N(u) by the transfer matrix. Let $w(u,r,\hat{t})$ represents the weight of tag $\hat{t}$ on the resource r given by the user u, and define as follows:

$$w(u,r,\hat{t}) = \frac{\sum_{v \in N(u) \wedge \hat{t} \in T(v)} sim(u,v)\, w(v,r,\hat{t})}{\sum_{u \in N(u) \wedge \hat{t} \in T(v)} sim(u,v)} \quad (12)$$

where N(u) is the nearest neighbor set of the user u, it includes the user u itself. Let $w(v,r,\hat{t})$ defines as the weight of tag $\hat{t}$ on the resource r given by the user v, and it can be get through the tag distribution on resources and personalized transfer matrix of the user v by Equation (13).

$$w(v,r,\hat{t}) = \mathbf{x}'_r \mathbf{y}_{v\hat{t}}^{\mathrm{T}} \quad (13)$$

After the transfer through the Equation (12), we can obtain the weight distribution of tags used by the user u and N(u). The tag vocabulary would be empty if there is no co-occurrence between tags on the resource r and the user u and N(u). In order to solve this problem, and increase the weight of tags that better represent the resource, we compute the final tag weight through Equation (14), and recommend the top n tags to the query user.

$$w(u,r,t) = \alpha w(u,r,t) + (1-\alpha) w(r,t) \tag{14}$$

where w(r, t) is the weight of tag t on the resource r. Parameter α uses to balance the weight distribution. When α equals 1.0, it considers the personalized transfer matrix and collaborative filtering technology together; when α equals 0.0, it is reduced to the popular tag recommendation method which only recommends the maximum tags on resources to the query user.

4. EXPERIMENTS AND ANALYSIS

4.1. Datasets

In order to prove the efficiency of our method, we conduct experiments on the Delicious dataset. Delicious is probably the best researched folksonomy to date [1]. Our dataset is a subset of the one presented in [22] , which is publicly available upon request. Here, we only keep the records in the first 5 months from September 2003 until January 2004, and ignore the time tags of the dataset. The experiments are conducted on p-core versions of the original graphs, with p set to 5. The p-core of a folksonomy graph is its largest subgraph where all users, items, and tags appear at least p times. The resulting global dataset statistics are shown in Table 2.

4.2. Evaluations

We use two standard measures from information retrieval to measure the recommendation effectiveness: precision and recall. Meanwhile, we test our method by F1 value, which is defined as the harmonic mean of precision and recall.

Table 2. Statistics for datasets and p-cores.

Dataset	p	Users	Tags	Resources	TAS	Bookmarks
Delicious (200309-200401)	1	3825	21.106	79.883	246.254	104.187
	5	397	478	805	13.361	6820

$$\text{Precision} = \frac{1}{N}\sum_{(u,r)\in \text{test}} \frac{\left|\{\text{recommend tags}\}\right| \cap \left|\{\text{actual annotation tags}\}\right|}{\left|\{\text{recommend tags}\}\right|}$$

$$\text{Recall} = \frac{1}{N}\sum_{(u,r)\in test} \frac{\left|\{\text{recommend tags}\}\right| \cap \left|\{\text{actual annotation tags}\}\right|}{\left|\{\text{actual annotation tags}\}\right|}$$

$$F1 = \frac{2 \times \text{Precision} \times \text{Recall}}{\text{Precision} + \text{Recall}}$$

where test refers to the test set, N is the number of the (u, r) in the test set.

4.3. Experiments

In this paper, we conduct the following experiments using different methods on the same dataset.

1) The traditional collaborative filtering method, denoted as colla. Here we adopt the algorithm described by Marinho [16] , for the

obtained nearest neighbor set N(u) of the query user u, the tag weight is calculated by Equation (15).

$$w(u,r,\hat{t}) = \sum_{v \in N(u)} sim(u,v)\delta(v,r,\hat{t}) \tag{15}$$

If $w(v,r,\hat{t}) \in A$, $\delta(v,r,\hat{t})$ equals 1.0, otherwise equals 0.0. Finally, we recommend the top n tags to the query user.

2) The transfer tensor of the user u based personalized tag recommendation, denoted as tensor u. This method is proposed by Wetzker [1] . It has been demonstrated that it can perform better than the improved Page Rank and Folk Rank algorighms and so on. The method first builds the transfer tensor between the personalized tags and folksonomies based on the tag co-occurrence, and further to mapping the tag distribution of resources to personalized tag vocabulary of the query user. Comparing with our method, it ignores the information of neighbors of the query user.

3) The method based on the transfer matrix and collaborative filtering technology, which proposed in this paper and denoted as trans + colla. It takes both the preference models of the query user u and his/her nearest neighbors into consideration.

The similarities between users in method (1) and (3) are calculated using three methods referred in subsection 3.3. There are denoted as DV which based on the resources, TV which based on the tags, and Trans V which based on the personalized transfer matrix, respectively.

4.4. Results

The experiments are conducted on 5-core dataset in Table 2. We perform 10-fold cross validation on the dataset by dividing it into 10 subsets randomly. The final evaluations of the results are the average mean of the 10-fold cross validation.

There are two parameters need to be addressed in the three recommend methods. One is the number of the nearest neighbors K, the other is the α in Equation (14). During the regulation, it finds that when K = 10 and $\alpha = 0.55$, method DV can obtain the best results; when K = 5 and $\alpha = 0.6$, method TV and TransV can get the best recommendations. When $\alpha = 0.6$, method tensor u can result the best recommendations.

We conduct the experiments with the above best parameter values, and the evaluations are shown in Figure 1, Figure 2 and Figure 3. As shown in Figure 2, the Recall is improving with the increase of recommend tags. Because evaluation Recall is the ratio of right recommend tags to the actual annotation tags, so the numerator of the Recall will increase if we recommend more tags. But on the other hand, the incorrect recommend tags will also increase, and this leads to the decrease of the Precision in Figure 1. It also indicates that the evaluation Recall and Precision restrict each other.

As shown in Figure 1, Figure 2 and Figure 3, method tensor_u performs better than the method colla_* (DV, TV, TransV), and method trans + colla_* also get superior results than colla_* as a whole. It indicates that the integrating of collaborative filtering approach is effective; the preference models of the query user's nearest neighbors also play an important role in the tag recommendation. Among the methods based on the transfer matrix, the evaluations finally tend to be the same with the increase of recommend tags, especially the Precision. The explanation here is the number of tags of a bookmark is small, that is to say, the number of tags assigned by a user to a resource is small, and the average is only

1.96. So it is important to recommend right tags when the number you can recommend is small.

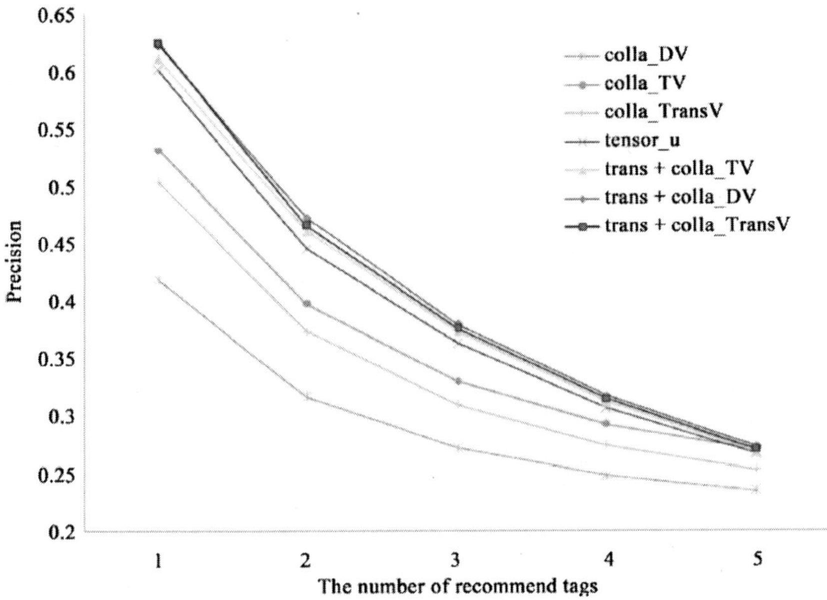

Figure 1. Results evaluation by precision.

In addition, we can find that the results of method trans + colla_* are similar. The similarity calculated by Equation (12) and Equation (13) is not the best. Because there are many users who share the same personalized tags with the query user due to the sparse dataset, and this lead to the higher similarity of transfer vectors between the query user and his/her nearest neighbor set. So due to the noise, it would produce some uninterested tags with higher tag weight, and finally influence the recommend results.

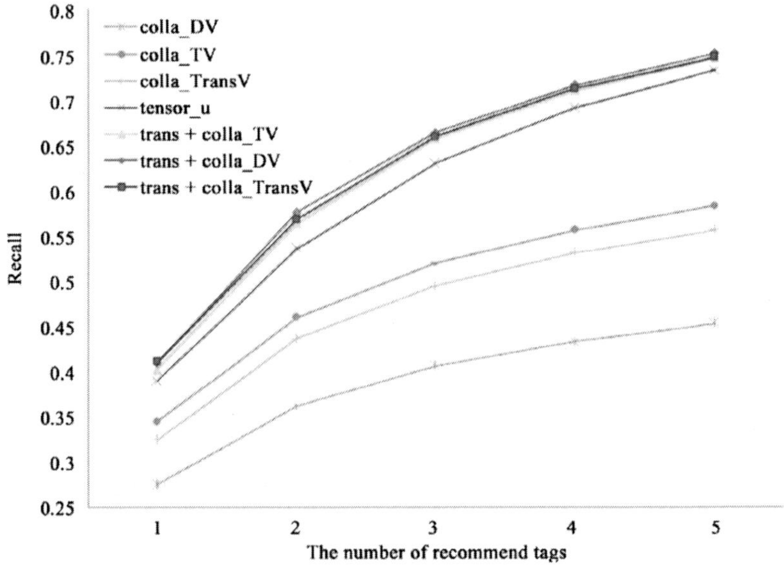

Figure 2. Results evaluation by recall.

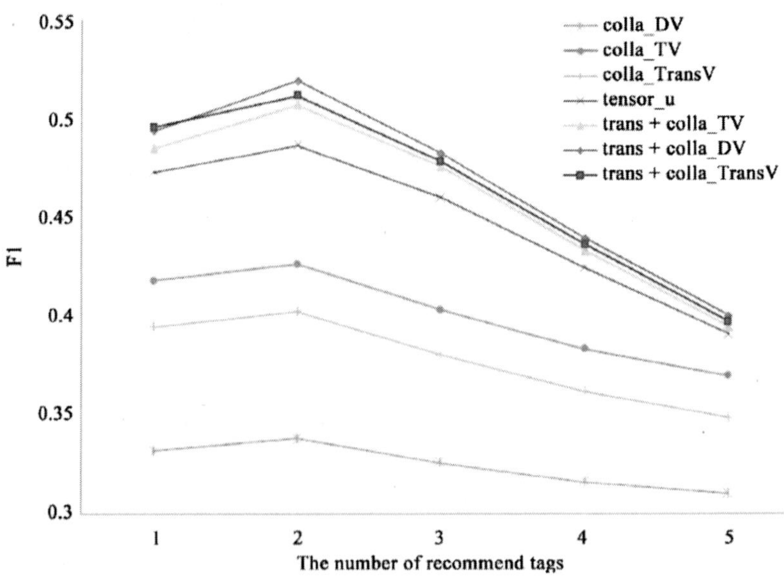

Figure 3. Results evaluation by F1 value.

5. CONCLUSIONS

Personalized tag recommendation can be better solved by transfer tensor, but this method limits to those used tags, and cannot find new interesting tags for users. So this paper integrates collaborative filtering technology to solve this problem. Firstly, we build the personalized transfer matrix for each user based on the co-occurrence of tags. The element of the matrix represents the mapping values between personalized tags and folksonomies. Secondly, combining the collaborative filtering approach, when recommend tags, we not only consider the preference model of query users, but also his/her nearest neighbors' and similarities between users. Aiming at the similarity, we also propose a method based on the personalized transfer matrix.

Further, we plan to focus on the problem of solving the redundant and ambiguous tags in the tagging system by applying the semantic models in the recommender system.

ACKNOWLEDGEMENTS

This work is partially supported by grant from the Natural Science Foundation of China (No. 61277370), Natural Science Foundation of Liaoning Province, China (No. 201202031, 2014020003), State Education Ministry and The Research Fund for the Doctoral Program of Higher Education (No. 20090041110002).

REFERENCES

1. Jäschke, R., Marinho, L., Hotho, A., Schmidt-Thieme, L. and Stumme, G. (2007) Tag Recommendations in Folksonomies.

Knowledge Discovery in Databases: PKDD 2007, Springer, Berlin, 506-514.

2. Wetzker, R., Zimmermann, C., Bauckhage, C. and Albayrak, S. (2010) I Tag, You Tag: Translating Tags for Advanced User Models. Proceedings of the 3rd ACM International Conference on Web Search and Data Mining, New York, 2010, 71-80.

3. Zhang, N., Zhang, Y. and Tang, J. (2009) A Tag Recommendation System for Folksonomy. Proceedings of the 2nd ACM Workshop on Social Web Search and Mining, New York, 2 November 2009, 9-16.

4. Guan, Z., Bu, J., Mei, Q., Chen, C. and Wang, C. (2009) Personalized Tag Recommendation Using Graph-Based Ranking on Multi-Type Interrelated Objects. Proceedings of the 32nd International ACM SIGIR Conference on Research and Development in Information Retrieval, New York, 19-23 July 2009, 540-547.

5. Symeonidis, P., Nanopoulos, A. and Manolopoulos, Y. (2008) Tag Recommendations Based on Tensor Dimensionality Reduction. Proceedings of the 2008 ACM Conference on Recommender Systems, New York, 23-25 October 2008, 43-50.

6. Rendle, S., Balby Marinho, L., Nanopoulos, A. and Schmidt-Thieme, L. (2009) Learning Optimal Ranking with Tensor Factorization for Tag Recommendation. Proceedings of the 15th ACM SIGKDD International Conference on Knowledge Discovery and Data Mining, New York, 2009, 727-736.

7. Rendle, S. and Schmidt-Thieme, L. (2010) Pairwise Interaction Tensor Factorization for Personalized Tag Recommendation. Proceedings of the 3rd ACM International Conference on Web Search and Data Mining, New York, 2010, 81-90.

8. Hotho, A., Jäschke, R., Schmitz, C. and Stumme, G. (2006) Bibsonomy: A Social Bookmark and Publication Sharing System. Proceedings of the Conceptual Structures Tool Interoperability Workshop at the 14th International Conference on Conceptual Structures, Aalborg, 2006, 87-102.

9. Hotho, A., Jäschke, R., Schmitz, C. and Stumme, G. (2006) Folkrank: A Ranking Algorithm for Folksonomies. Vol. 1, In: Althoff, K.D., Ed., Klaus-Dieter Althoff, LWA, Hildesheim, 111-114.

10. Hotho, A., Jäschke, R., Schmitz, C. and Stumme, G. (2006) Information Retrieval in Folksonomies: Search and Ranking. Springer, Berlin, 411-426.

11. Sigurbjörnsson, B. and Van Zwol, R. (2008) Flickr Tag Recommendation Based on Collective Knowledge. Proceedings of the 17th International Conference on World Wide Web, Beijing, 21-25 April 2008, 327-336.

12. Chen, W.Y., Zhang, D. and Chang, E.Y. (2008) Combinational Collaborative Filtering for Personalized Community Recommendation. Proceedings of the 14th ACM SIGKDD International Conference on Knowledge Discovery and Data Mining, Las Vegas, 24-27 August 2008, 115-123.

13. Golder, S.A. and Huberman, B.A. (2006) Usage Patterns of Collaborative Tagging Systems. Journal of Information Science, 32, 198-208.

14. Marlow, C., Naaman, M., Boyd, D. and Davis, M. (2006) HT06, Tagging Paper, Taxonomy, Flickr, Academic Article, to Read. Proceedings of the Seventeenth Conference on Hypertext and Hypermedia, Odense, 22-25 August 2006, 31- 40.

15. Mika, P. (2005) Ontologies Are Us: A Unified Model of Social Networks and Semantics. In: Gil, Y., Motta, E., Benjamins, V.R.

and Musen, M.A., Eds., The Semantic Web-ISWC 2005, Springer, Berlin Heidelberg, 522-536.

16. Marinho, L.B. and Schmidt-Thieme, L. (2008) Collaborative Tag Recommendations. In: Preisach, C., Burkhardt, H., Schmidt-Thieme, L. and Decker, R., Eds., Data Analysis, Machine Learning and Applications, Springer, Berlin Heidelberg, 533-540.

17. Xu, Z., Fu, Y., Mao, J. and Su, D. (2006) Towards the Semantic Web: Collaborative Tag Suggestions. Proceedings of the Collaborative Web Tagging Workshop at WWW, Edinburgh, 23-26 May 2006, 1-8.

18. Koren, Y. (2010) Collaborative Filtering with Temporal Dynamics. Communications of the ACM, 53, 89-97.

19. Gueye, M., Abdessalem, T. and Naacke, H. (2014) A Parameter-Free Algorithm for an Optimized Tag Recommendation List Size. Proceedings of the 8th ACM Conference on Recommender Systems, ACM Press, New York, 233-240.

20. Gueye, M., Abdessalem, T. and Naacke, H. (2013) An Efficient Trust-and Popularity-Based Tag Recommender.

21. Ifada, N. and Nayak, R. (2015) Do-Rank: DCG Optimization for Learning-to-Rank in Tag-Based Item Recommendation Systems. In: Cao, T., Lim, E.-P., Zhou, Z.-H., Ho, T.-B., Cheung, D. and Motoda, H., Eds., Advances in Knowledge Discovery and Data Mining, Springer International Publishing, Berlin Heidelberg, 510-521.

22. Wetzker, R., Zimmermann, C. and Bauckhage, C. (2008) Analyzing Social Bookmarking Systems: A Delicious Cookbook. Proceedings of the ECAI 2008 Mining Social Data Workshop, IOS Press, Amsterdam, 26-30.

CHAPTER 5

Time Series Prediction Method of Bank Cash Flow and Simulation Comparison

Wen-Hua Cui, Jie-Sheng Wang * and Chen-Xu Ning

School of Electronic and Information Engineering, University of Science and Technology Liaoning, Anshan 114044, Liaoning, China

ABSTRACT

In order to improve the accuracy of all kinds of information in the cash business and enhance the linkage between cash inventory forecasting and cash management information in the commercial bank, the first moving average prediction method, the second moving average prediction method, the first exponential smoothing prediction and the second exponential smoothing prediction methods are adopted to realize the time series prediction of bank cash flow, respectively. The prediction accuracy of the cash flow time series is improved by optimizing the algorithm parameters. The simulation experiments are carried out on the reality commercial bank's cash flow data and the predictive performance comparison results show the effectiveness of the proposed methods.

KEYWORDS

time series prediction; moving average prediction; exponential smoothing prediction

1. INTRODUCTION

In recent years, the time series modeling and prediction have been one of the most active research topics in the academic research and engineering practice [1,2,3,4]. The time series is usually a chronological series of observed data (information) according to the time sequence, whose values are sampled at the invariable time intervals. Researchers often predict future changes based on the historical data. For example, according to the situation in the past or the current period of the market sales, the changes of stock prices, the population growth and the bank's deposit and withdrawal, the changes of the market sales, the changes of stock prices, the population growth and the bank's deposit and withdrawal in the future are predicted. The time series forecasting affects the life of people everywhere, so it has an important practical significance and research prospects in every field of today's society, which is also an important direction in the computer application field [5,6,7,8].

The bank cash flow forecasting management information system is designed to create a system management platform for the prediction and analysis of the commercial bank cash flow. It will realize the cash flow data statistics summary, the cash flow short-term and long-term prediction, the summary and statistical analysis comprehensively and scientifically of the business information, the operational information and the management information related to the commercial bank cash flow under three levels: secondary branches (Cash Operation Center), branch (Business Library) and Network. Its purpose is to provide effective data all levels of organization to analyze and assess cash business operation conditions. Also it will provide effective system management means for the cash operation managers and decision-making people at all levels.

Aiming at the existed problem in the analysis of commercial bank cash flow, four time series prediction methods are used to set up the prediction models. The simulation results show the effectiveness of the proposed methods. The paper is organized as follows. In Section 2, the time series prediction methods of bank cash flow are introduced. The simulation experiments and results analysis are introduced in details in Section 3. Finally, the conclusion illustrates the last part.

2. TIME SERIES PREDICTION METHODS OF BANK CASH FLOW

The time series prediction method analyzes the predicted target's changes with the time on the times series composed of the historical data according to the chronological order. The prediction method is quantitative and the related mathematical model is established for extrapolation [9,10,11,12,13,14,15]. Based on the circulation data of bank cash, four prediction methods of the various cash operating specific modules are designed for providing the decision-making basis of the business plan in commercial bank. The flowchart of the time series prediction methods of bank cash flow is shown in Figure 1. The simulation environment is described as follows: *Windows* 7 operating system, *Intel* processor (2.5 GHz, 4G memory) and Matlab 2010 simulation software.

2.1. The First Moving Average Predictive Method

The moving average method is a predictive technique developed on the basis of the arithmetic average. The arithmetic average method can only reflect the average of a set of data and cannot reflect the change trend of the data. However, the basic idea of the moving average method is item by item based on time series data by taking a certain number of cycles of data to be averaged every time, successive advancing in chronological order. In each propulsion cycle, the forward cycle data is rejected, a new cycle data is added and the average value is calculated [16]. Set X_t is the actual value of the cycle t. The first moving average value is calculated by the following equation.

$$M_t^{(1)}(N)=\frac{X_t+X_{t-1}+\ldots+X_{t-N+1}}{N}=\sum_{i=0}^{N-1}X_{t-i}/N \tag{1}$$

So the predictive value of the next period (t + 1) is:

$$\hat{X}_{t+1}=M_t^{(1)} \tag{2}$$

where N is the number of selected data for calculating the first moving average value and X^t+1 is the predictive value of the next period (t + 1).

Based on the collected data in a commercial bank from January to April in 2011and 2012, the first moving average method is adopted to realize the time series prediction of bank cash flow. The cycle number of the moving average in 2011 is 2–15 and the continuous sample points are 100. The optimal moving average value for the selected data is N = 6 and the simulation results are shown in Figure 2. The cycle number of the moving average in 2012 is 2–15 and the continuous sample points are 100. The optimal moving average value for the selected data is N = 2 and the simulation results are shown in Figure 3.

It can be seen from Figure 2 and Figure 3 that the random fluctuations of bank actual cash flow values is larger, but the random fluctuations are reduced significantly after the first moving average calculations. The more used months by calculating the average, that is to say the greater N, the stronger of the smoothing degree and the more small fluctuation. However, in this case, the reaction velocity on this change trend for the actual data is slower. Conversely, if N is set lower, the reaction velocity on this change trend for the actual data is more sensitive, but the smoothing degree is less and it is easy to reflect the random interference as the trend. Therefore, the choice of N is very important, which should be selected according to the specific situation.

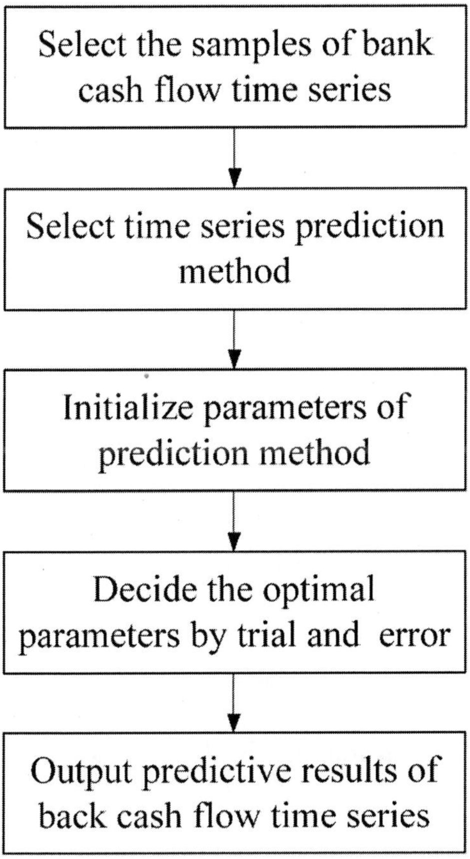

Figure 1. Flowchart of the time series prediction methods of bank cash flow.

When N is equal to the cycle of change, the influence of periodic change can be eliminated. In practice, the mean square error S of prediction for past data is generally adopted to be the criterion for choosing N. However, because of the accumulation of errors, the predictive error is bigger for the prediction of more distant periods. Therefore, the first moving average method is usually adapted to the stationary model and only a period prediction (*i.e.*, predicting the $t + 1$ period). So when the basic mode of predicted variables changes, the adaptability of the first moving average method will be relatively poor.

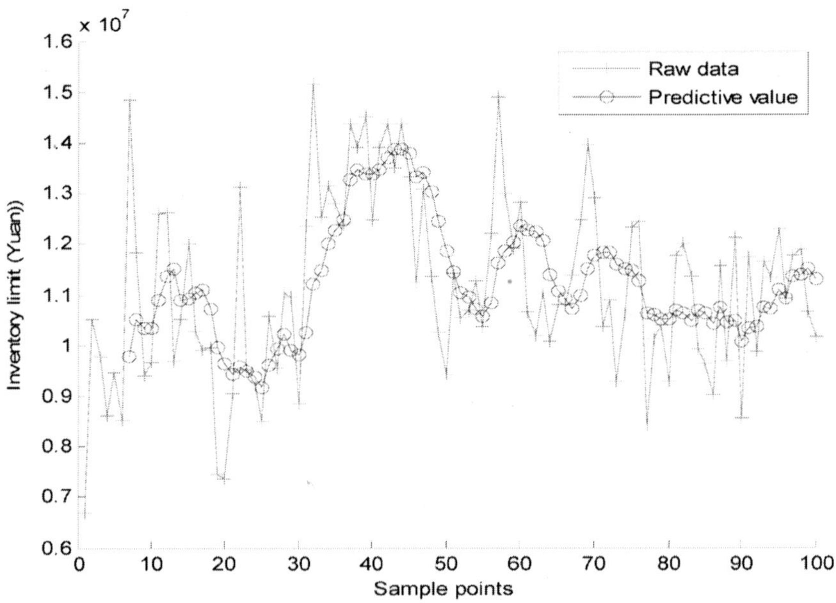

Figure 2. Predictive results of the first moving average method in 2011.

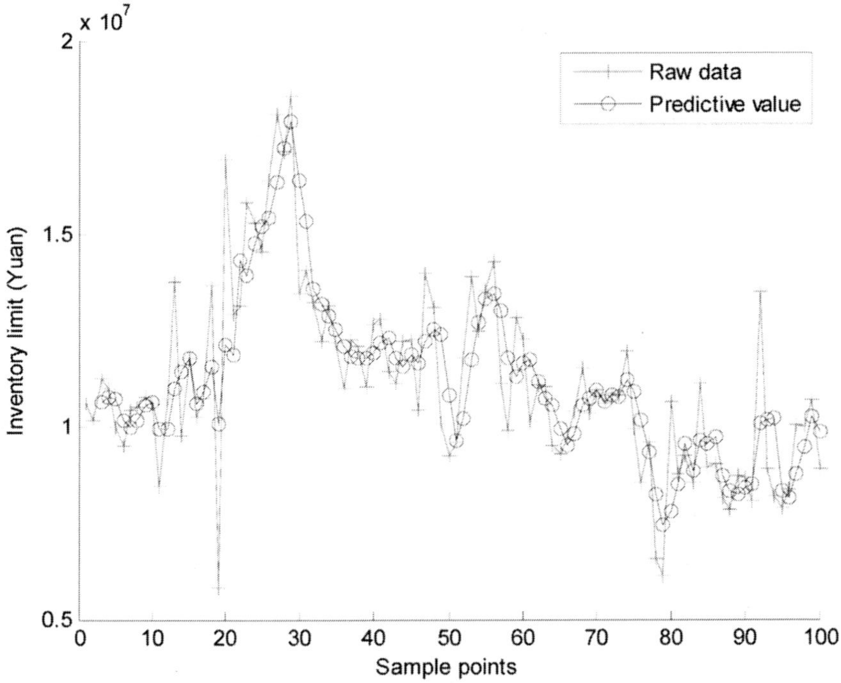

Figure 3. Predictive results of the first moving average method in 2012.

2.2. The Second Moving Average Predictive Method

The second moving average method is to carry out the moving average again based on the first moving average method, that is to say the prediction model is established on the basis of the first moving average values and the second moving average values to calculate the predicted value. As mentioned above, the average value calculated by the first moving average method has a lag deviation. Especially when there is a linear trend in the time series data, the first moving average value is always behind the observation value. The second moving average method is proposed to revise this lag deviation by establishing the mathematical model having the linear time relationship of the predictive target to obtain the predictive values.

The second moving average predictive method can solve the contradiction of the predicted value lagging behind the actual observed values, which are suitable for forecasting the time series with the phenomenon having the obvious change trend. At the same time, it also retains the advantage of the first moving average method. Set $M_t^{(2)}$ is the second moving average value in period t, Y_t is the actual value of the cycle t. The second moving average method is described as follows.

$$M_t^{(1)} = \frac{Y_t + Y_{t-1} + \ldots + Y_{t-n+1}}{n} \tag{3}$$

$$M_t^{(2)} = \frac{M_t^{(1)} + M_{t-1}^{(1)} + \ldots M_{t-n+1}^{(1)}}{n} \tag{4}$$

where $M_t^{(1)}$ is the first moving average value in period t, $M_t^{(2)}$ is the second moving average value in period t, n is the number of selected data for calculating the first moving average value.

In order to eliminate the influence of lag error on the prediction results, the linear trend model with the hysteresis error rule on the basis of the first and second moving average values is established. $M_t^{(1)}$ and $M_t^{(2)}$ are used to estimate the intercept aˆt and slope bˆt of the linear trend model. The calculation formulas are described as follows.

$$\begin{cases} \hat{a}_t = 2M_t^{(1)} - M_t^{(2)} \\ \hat{b}_t = \dfrac{2}{N-1}(M_t^{(1)} - M_t^{(2)}) \end{cases} \tag{5}$$

Then the linear trend prediction model is established as:

$$\hat{y}_{t+\tau} = \hat{a}_t + \hat{b}_t \tau \tag{6}$$

where t is the current period, τ is from t to forecast period number, y^t+τ is predictive value in period $t + \tau$, a^t is the estimated value of the intercept, b^t is the estimated value of the slope.

In order to carry out the research on the second moving average method for predicting the bank cash flow time series, the same data in the first moving average method are adopted. The cycle number of the moving average in 2011 and 2012 is 2–15 and the optimal moving average value for the selected data is N = 3. The simulation results are shown in Figure 4 andFigure 5. It can be seen from Figure 4 and Figure 5 that the predicted effect of the second moving average method is better than the first moving average method. So it is more suitable for the prediction the time series with linear trend changes.

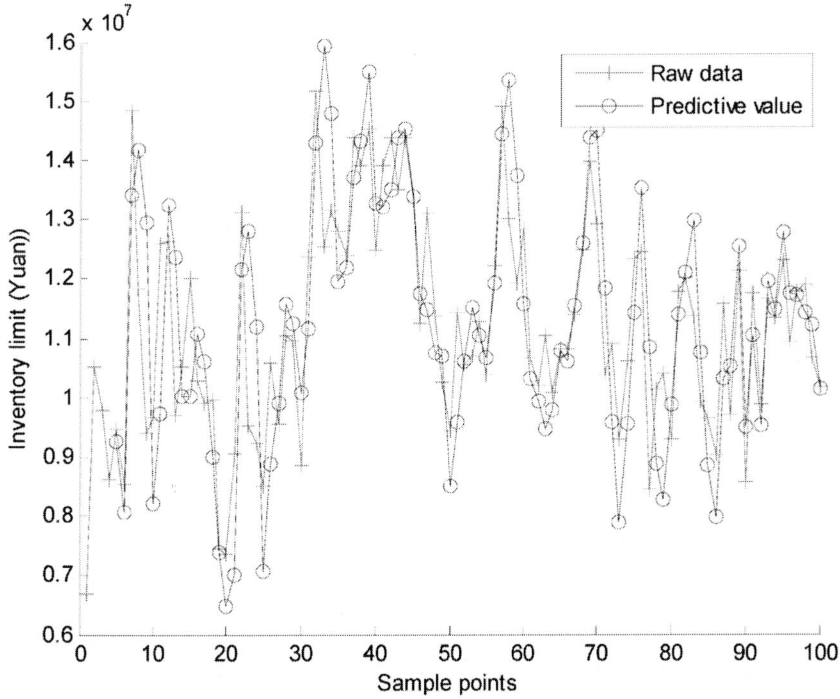

Figure 4. Predictive results of the second moving average method in 2011.

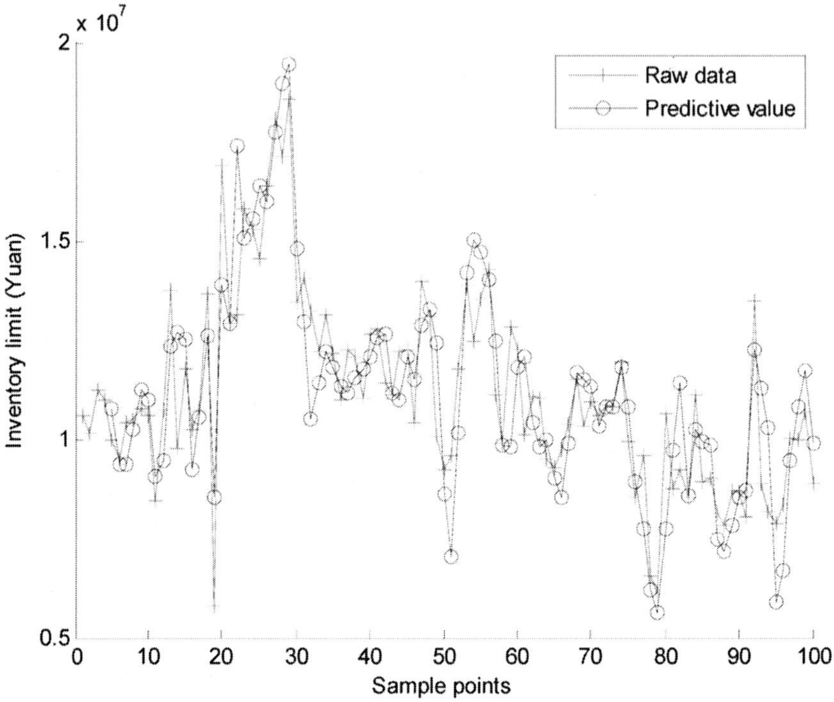

Figure 5. Predictive results of the second moving average method in 2012.

2.3. The First Exponential Smoothing Prediction Method

The exponential smoothing method is a time sequence analysis prediction method developed from the moving average method. By calculating the exponential smoothing value, the future of phenomenon is predicted with the certain time series prediction model. Its principle is to calculate the exponential smoothing values in any period with the style of weighted average by combining the actual observed value and the exponential smoothing value in the previous period. The first exponential smoothing method is a weighted prediction, whose weight coefficient is α. It need not store all history data, which can greatly reduce the data storage.

Sometimes, only a new observation value, the latest predicted value and the weight coefficient α can be used to realize the time series prediction.

When the time series has no obvious trend change, the first exponential smoothing prediction method is effective. Set $X_0, X_1, \ldots, X_n$ are the observations of time series and $S_1^{(1)}, S_2^{(1)}, \ldots, S_n^{(1)}$ are the exponential smoothing values of observations in time t. So the first exponential smoothing value is calculated by the following equation.

$$S_t^{(1)} = \alpha X_t + \alpha(1-\alpha)X_{t-1} + \alpha(1-\alpha)^2 X_{t-2} + \ldots \tag{7}$$

where α is smoothing coefficient, $0 < \alpha < 1$.

It can be seen form Equation (7) that the weight coefficients of the actual values X_t, X_{t-1} and X_{t-2} are α, $\alpha \cdot (1 - \alpha)$ and $\alpha \cdot (1 - \alpha)^2$, respectively. The data is farther from the moment, with the smaller weight coefficient. Because the weight coefficient is an index geometric series, it is called the exponential smoothing method. Equation (7) can be changed slightly as:

$$\begin{aligned} S_t^{(1)} &= \alpha X_t + (1-\alpha)\left[\alpha X_{t-1} + \alpha(1-\alpha)X_{t-2} + \ldots\right] \\ &= \alpha X_t + (1-\alpha)S_{t-1}^{(1)} \end{aligned} \tag{8}$$

Equation (8) can be rewritten as:

$$S_t^{(1)} = S_{t-1}^{(1)} + \alpha(X_t - S_{t-1}^{(1)}) \tag{9}$$

Then the predictive value of next cycle is obtained:

$$\hat{X}_{t+1} = S_t^{(1)} \tag{10}$$

$$\hat{X}_{t+1} = \hat{X}_t + \alpha(X_t - \hat{X}_t) \tag{11}$$

Seen from the Equation (11), the exponential smoothing method solves a problem existing in the moving average method,*i.e.*, threre is no longer a need to store the historical data in past N cycles. One only needs the most recent observed value x_t, the most recent predicted value X^t and the weight coefficient α to calculate a new predictive value.

The exponential smoothing method is used to realize the unequal weights data handle with different times by using the smoothing coefficient α. So the problem of chosing α is discuss as follows. (1) $0 < \alpha < 1$. (2) The selection of the smoothing coefficient α is an important problem, which affects the prediction results directly. It is selected according to the characteristics of the actual time series and expert experiences.

1) When the fluctuation of time series is not big and the long-term trend is relatively stable, it is better to select a smaller α. The weight of the past predictive values can then be aggravated. Its value generally locates the scope [0.05, 0.2].

2) When the fluctuation of time series is big and has an obvious tendency of rapid change, it is better to select the big α, which can aggravate the weight of the new predictive values. Its value generally locates the scope [0.3, 0.7].

In order to carry out the research on first exponential smoothing prediction method for predicting the bank cash flow time series, the same data in the above two methods are adopted. When the weight coefficient is 0.9, the mean square error is minimum and the simulation results are shown in Figure 6 and Figure 7. It can be seen from Figure 6 and Figure 7 that the first exponential smoothing model prediction method need to find the best value α through the repeated experiments. However, even with a best value α, the accuracy of predicted results also is not ideal for the time series with larger trend.

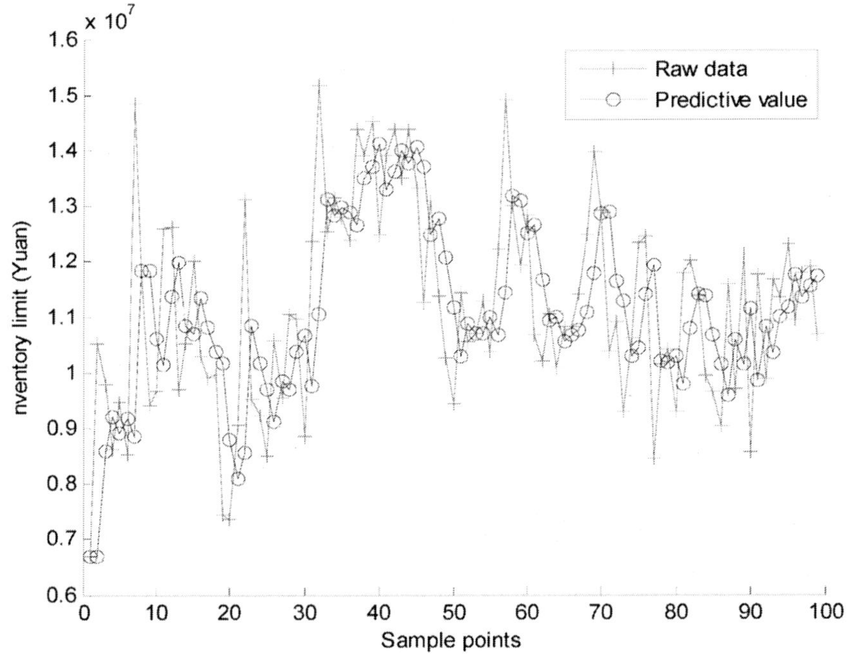

Figure 6. Predictive results of the first exponential smoothing prediction method in 2011.

2.4. The Second Exponential Smoothing Prediction Method

Because the first moving average method and second linear moving average method all need store a large amount of historical data, in order to compensate for this limitation, the second linear exponential smoothing model is developed, which only need to use three data values and a smoothing coefficient α. At the same time, it also can make the weight of past observations small. In most cases, the use of second linear exponential smoothing model is more convenient.

The second exponential smoothing prediction method is to carry out another exponential smoothing based on the first exponential smoothing prediction method, which has the advantages of simple calculation, less sample requirement, strong adaptability and the relatively stable predicted results. It cannot predict lonely and must cooperate with the first exponential smoothing prediction method to build the time series predictive mathematical model. The weighted average of historical data is as the predicted value of future time. The second exponential smoothing prediction method is described as follows.

$$S_t^{(1)} = aY_t(1-\mathrm{a})\mathrm{S}_{(t-1)}^{(1)} \tag{12}$$

$$S_t^{(2)} = aS_t^{(1)} + (1-\mathrm{a})\mathrm{S}_{(t-1)}^{(2)} \tag{13}$$

where, $S_t^{(1)}$ is the first exponential smoothing value in period t, $S_t^{(2)}$ is the second exponential smoothing value in period $t$and α is the smoothing constant. So the prediction model of the second exponential smoothing method is described as follows.

$$F_{t+T} = a_t + b_t T \tag{14}$$

$$a_t = 2S_t^{(1)} - S_t^{(2)} \tag{15}$$

$$b_t = (a/1-\mathrm{a})(\mathrm{S}_t^{(1)} - \mathrm{S}_t^{(2)}) \tag{16}$$

where F_{t+T} is the predictive value in period $t + T$, T is the number of the future forecast period, a_t and b_t are the model parameters.

Equation (15) is used to add the difference between the first exponential smoothing value and the second exponential smoothing value on the first exponential smoothing value. Equation (16) is adopted to add the trend change value. When predicting the period $t + 1$, a trend change value b_t is added to the a_t. When predicting the period $t + T$, the trend change value $T \times b_t$ is added to the a_t.

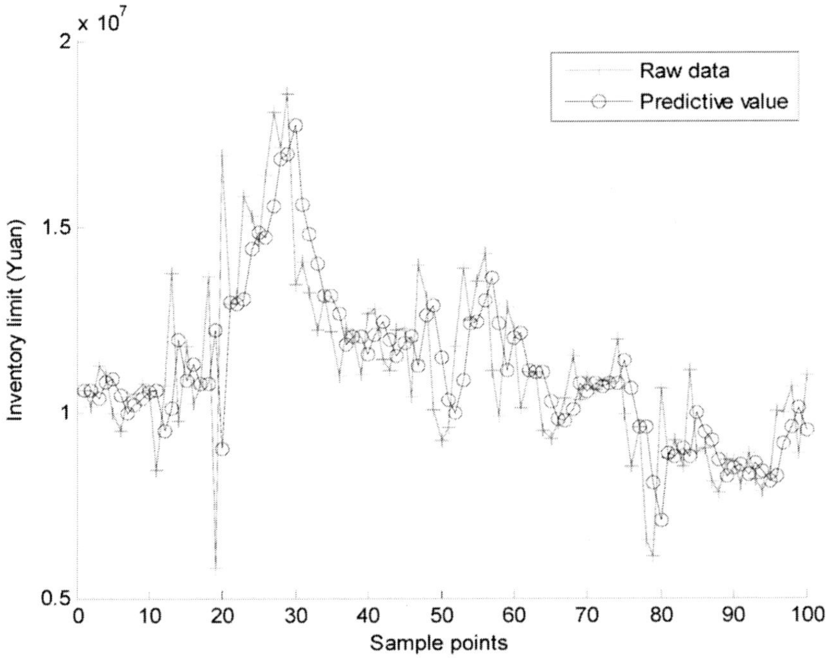

Figure 7. Predictive results of the first exponential smoothing prediction method in 2012.

In order to carry out the research on the second exponential smoothing prediction method for predicting the bank cash flow time series, the same data in the above three methods are adopted. When the weight coefficient is 0.5, the mean square error is minimum and the simulation results are shown in Figure 8 and Figure 9.

It can be seen from Figure 8 and Figure 9 that because the second exponential smoothing method considers the linear parameter changes in different periods of the time series, the extent of the predicted values fitting with the original time series is very good, which reflects the variation trends of the original time series in different time periods. So it is more suitable for the prediction of time series with a linear trend.

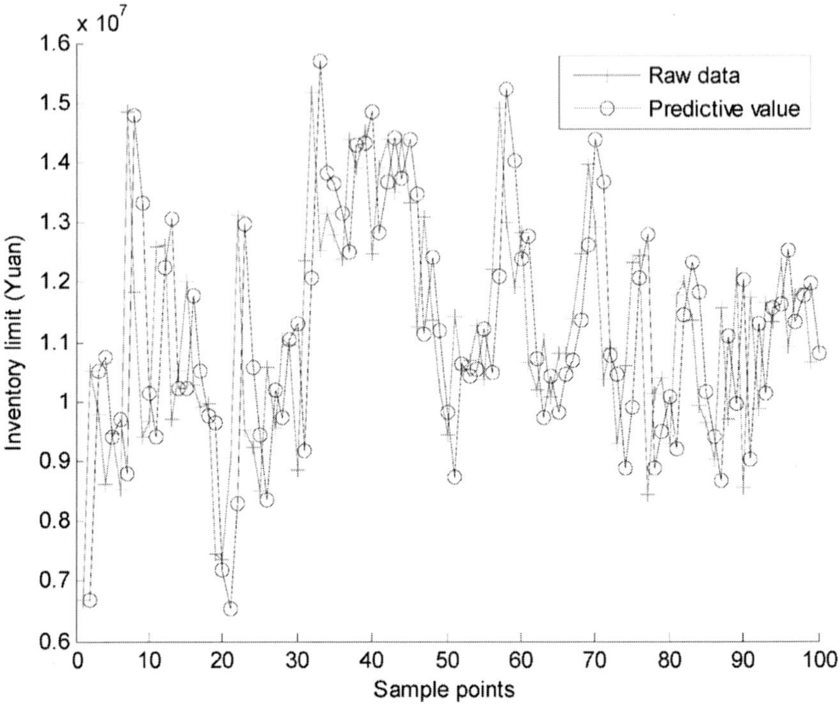

Figure 8. Predictive results of the second exponential smoothing method in 2011.

3. SIMULATION COMPARISON

In order to carry out the performance comparison under the above mentioned four time series predictive methods for bank cash flow, the simulation curves are shown in Figure 10. It can be seen clearly that the predicted results of the second exponential smoothing method is better than other methods.

The prediction error is the deviation between the predicted results and the actual results, which determines the prediction accuracy. The accuracy of the quantitative prediction methods has a lot of measurable indicators. Suppose $y_1, y_2, \ldots, y_n$ are the actual observed values of the predicted target and y^1,y^2,…,y^n are the predicted values.

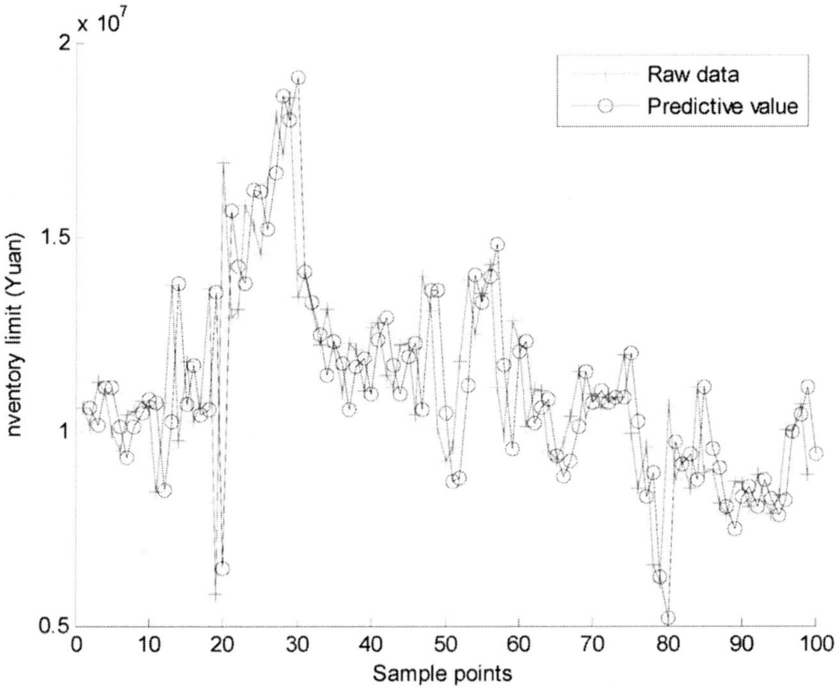

Figure 9. Predictive results of the second exponential smoothing method in 2012.

The absolute error of predicted points is

$$a_t = y_t - \hat{y}_t, \quad t = 1, 2, \cdots, n \tag{17}$$

where a_t is the absolute error at the point t. Obviously, a_t is the most direct measure index of prediction error, but it is affected by the measurement unit of predicted object. So it is unsuitable as the final measure indicator of prediction accuracy.

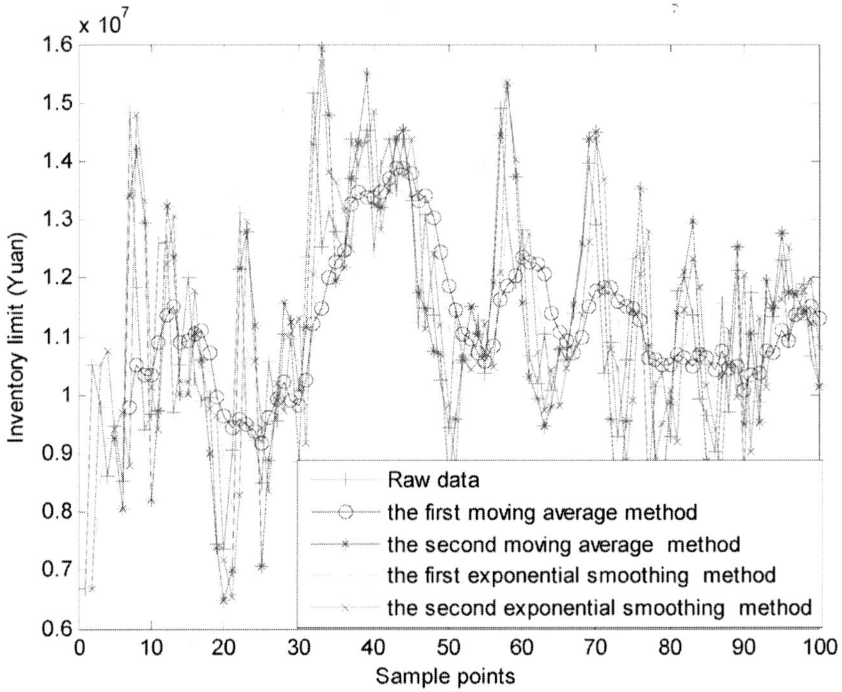

Figure 10. Performance comparison results.

Relative error of predicted points

$$\hat{a}_t = \frac{a_t}{y_t} = \frac{y_t - \hat{y}_t}{y_t}, \quad t = 1, 2, \cdots, n \tag{18}$$

where a^t is the relative error at the point *t*, which is usually expressed as a percentage and measures the accuracy of the predicted values relative to the observed value at the predicted point *t*.

Prediction accuracy of the prediction points

$$A_t = 1 - \left| y_t - \hat{y}_t \right| / y_t \, , \; 0 \le \left| y_t - \hat{y}_t / y_t \right| \le 1 \tag{19}$$

$$A_t = 0,\ |y_t - \hat{y}_t|/y_t > 1 \tag{20}$$

where A_t is the prediction accuracy at the prediction point t.

Mean square error (MSE)

Mean square error (MSE) is a kind of convenient method to measure the average error to evaluate the degree of data change, which is described as follows.

$$MSE = \frac{1}{n}\sum_{t=1}^{n}(y_t - \hat{y}_t)^2 \tag{21}$$

Four mentioned methods are used to carry out the simulation on the same pretreated experimental. The simulation results are shown in Table 1. The performance comparison results in Table 1 also showed that the second exponential smoothing method obtains the higher prediction precision of the cash flow time series.

4. CONCLUSIONS

Four time series predictive methods are adopted to realize the real-time prediction of cash flow in the commercial bank. By comparing the time series predictive performance under four algorithms and analyzing the simulation results in-depth, the second exponential smoothing prediction method is the optimal prediction method.

Table 1. Performance comparison results.

Performance indicators	First moving average method	Second moving average method	First exponential smoothing method	Second exponential smoothing method
MSE	2.47×10^{12}	1.55×10^{12}	2.67×10^{12}	1.02×10^{12}
Absolute error	1.14×10^{6}	6.23×10^{5}	1.06×10^{6}	1.12×10^{5}
Relative error of (%)	5.2	2.9	3.4	1.1
Prediction accuracy (%)	94.8	97.1	96.6	98.9

ACKNOWLEDGMENTS

This work is partially supported by the Program for China Postdoctoral Science Foundation (Grant No. 20110491510), the Program for Liaoning Excellent Talents in University (Grant No. LR2014008) and the Project by Liaoning Provincial Natural Science Foundation of China (Grant No. 2014020177).

AUTHOR CONTRIBUTIONS

Wen-Hua Cui participated in the concept, design, interpretation and commented on the manuscript. A substantial amount of Jie-Sheng Wang's contribution to the draft writing and critical revision of this paper was undertaken. Chen-Xu Ning participated in the data collection, analysis and algorithm simulation.

CONFLICTS OF INTEREST

The authors declare no conflict of interest.

REFERENCES

1. Bar-Joseph, Z. Analyzing time series gene expression data. *Bioinformatics* **2004**, *20*, 2493–2503.

2. Dong, Z. Study on the time-series modeling of China's per capita GDP. *Contemp. Manag.* **2006**, *11*, 15.

3. Kasabov, N.K.; Song, Q. DENFIS: Dynamic evolving neural-fuzzy inference system and its application for time-series prediction. *IEEE Trans. Fuzzy Syst.* **2002**, *10*, 144–154.

4. Martinetz, T.M.; Berkovich, S.G.; Schulten, K.J. Neural-gas' network for vector quantization and its application to time-series prediction. *IEEE Trans. Neural Netw.* **1993**, *4*, 558–569.

5. Chen, Y.; Yang, B.; Dong, J. Time-series prediction using a local linear wavelet neural network. *Neurocomputing* **2006**,*69*, 449–465.

6. Xie, J.X.; Cheng, C.T.; Chau, K.W.; Pei, Y.Z. A hybrid adaptive time-delay neural network model for multi-step-ahead prediction of sunspot activity. *Int. J. Environ. Pollut.* **2006**, *28*, 364–381.

7. Shi, Z.; Han, M. Ridge regression learning in ESN for chaotic time series prediction. *Control Decis.* **2007**, *22*, 258.

8. Li, D.; Han, M.; Wang, J. Chaotic time series prediction based on a novel robust echo state network. *IEEE Trans. Neural Netw. Learn. Syst.* **2012**, *23*, 787–799.

9. Long, W.; Liang, X.M.; Long, Z.Q.; Qin, H.Y. RBF neural network time series forecasting based on hybrid evolutionary algorithm. *Control Decis.* **2012**, *8*, 029.

10. Zhang, G.P. Time series forecasting using a hybrid ARIMA and neural network model. *Neurocomputing* **2003**, *50*, 157–175.

11. Silva, C.G. Time series forecasting with a nonlinear model and the scatter search meta-heuristic. *Inf. Sci.* **2008**, *178*, 3288–3299.

12. Sun, P.; Marko, K. The square root Kalman filters training of recurrent neural networks. In Proceedings of IEEE International Conference on Systems, Man, and Cybernetics, San Diego, CA, USA, 11–14 October 1998.

13. Martinetz, T.M.; Berkovich, S.G.; Schulten, K.J. Neural-gas' network for vector quantization and its application to time-series prediction. *IEEE Trans. Neural Netw.* **1993**, *4*, 558–569.

14. Connor, J.T.; Martin, R.D.; Atlas, L.E. Recurrent neural networks and robust time series prediction. *IEEE Trans. Neural Netw.* **1994**, *5*, 240–254.

15. Van Gestel, T.; Suykens, J.A.K.; Baestaens, D.-E.; Lambrechts, A.; Lanckriet, G.; Vandaele, B.; de Moor, B.; Vandewalle, J. Financial time series prediction using least squares support vector machines within the evidence framework. *IEEE Trans. Neural Netw.* **2001**, *12*, 809–821.

16. Aras, C.M.; Jeffrey, D.M.; Roderick, K.S. System for allocation of network resources using an autoregressive integrated moving average method. U.S. Patent No. 5,884,037, 16 March 1999.

CHAPTER 6

A Smoothing Neural Network Algorithm for Absolute Value Equations

Feiran Wang, Zhensheng Yu, Chang Gao

College of Science, University of Shanghai for Science and Technology, Shanghai

ABSTRACT

In this paper, we give a smoothing neural network algorithm for absolute value equations (AVE). By using smoothing function, we reformulate the AVE as a differentiable unconstrained optimization and we establish a steep descent method to solve it. We prove the stability and the equilibrium state of the neural network to be a solution of the AVE. The numerical tests show the efficient of the proposed algorithm.

KEYWORDS

Absolute Value Equations, Neural Network, Smoothing Function, Linear Complementarity Problem

1. INTRODUCTION

Consider the following absolute value problem [1] -[3] :

$$Ax-|x|=0 \tag{1}$$

where $A \in R^{n\times n}, x,b \in R$, $|x|$ is absolute value of x, it is a subclass of absolute value equations $Ax-B|x|=b$ which is proposed by Rohn [4] , and it is a NP-hard problem [1] .

The AVE has closed relation with some important problems, for example, the linear programming, Quadratic programming problem and the bimatrix game problem. The above problems can be transformed into the linear complementarity problem, and the linear complementarity problem can be transformed into the absolute value equations. Due to its simple and special structure and application value, the research on absolute value equation has drawn attention of many researchers. Mangasarian [5] pointed out the relationship between backpack feasibility problem and the AVE. The problem of AVE has been studied deeply by Yamashita and Fukushima [6] , and the results of the research on the problem of AVE are applied to the problem of location selection, good results are obtained. The numerical solution methods of AVE, such as Newton method, quasi-Newton method, are reachable in [7] -[12] .

In this paper, we present a smooth approximation function which is based on neural network method to solve the AVE. By using a smooth approximation function of $|x|$, we turn it into a differentiable unconstrained optimization problem. Furthermore, we obtain the approximate solution of the original problem based on our established unconstrained optimization problem and the neural network model. Compared with the Newton method, the neural network model needs less requirement for the hardware of compute and the iterative process is real-time.

2. THE SMOOTHING REFORMULATING OF AVE

The absolute value Equation (1) is equivalent to the nonlinear equations:

$$\Theta(x)=0 \tag{2}$$

where $\Theta(x)=Ax-|x|-b$. Since it is a non smooth function, we construct a smooth function to approximate it.

Definition 1.1 Smoothing approximation function, given a function $\Theta:=R^n\rightarrow R^n$, smoothing function $\Theta_\mu:=R^n\rightarrow R^n\,(\mu>0)$ is called smoothing approximation function, if for any $x\in R^n$, there exists $\kappa>0$ so that

$$\left\|\Theta_\mu(\cdot)-\Theta(\cdot)\right\|\leq\kappa\mu,\quad\forall\mu>0$$

where κ is not dependent on the x.

In this paper, we use the aggregate function [13] to give a smooth approximation of the absolute value equation:

Let $\phi(x)=|x|$, so every component of the absolute value function can be recorded as

$$\phi(x_i) = |x_i| = \max\{x_i, -x_i\}, \quad i = 1, 2, \cdots, n$$

For any $\phi(x_i)$, the definition of smoothing function is as follows

$$\phi_\mu(x_i) = \mu \ln\left(\exp\left(\frac{x_i}{\mu}\right) + \exp\left(-\frac{x_i}{\mu}\right)\right), \quad i = 1, 2, \cdots, n$$

So the function of absolute value $|x|$ is obtained as follows

$$\phi_\mu(x) = \left(\phi_\mu(x_1), \phi_\mu(x_2), \cdots, \phi_\mu(x_n)\right)^{\mathrm{T}}$$

Thus the absolute value equation is transformed into the following smooth equations

$$\Theta_\mu(x) = 0 \tag{3}$$

where $\Theta_\mu(x) = Ax - \phi_\mu(x) - b$

We define the function as follows

$$\Psi_\mu(x) = \frac{1}{2}\|\Theta_\mu(x)\|^2$$

where $\Psi_\mu(x)$ is the smoothing approximation of $\Psi(x) = \frac{1}{2}\|\Theta(x)\|^2$, $\Psi_\mu(x)$ is said as the energy function

of the neural network. Thus, the approximate solution of the absolute value equation is transformed to the global optimal solution of the optimization problem $\min \Psi_\mu(x)$

3. NEURAL NETWORK MODEL FOR ABSOLUTE VALUE EQUATION

Consider the following unconstrained optimization problem

$$\Psi_\mu(x)=\frac{1}{2}\|\Theta_\mu(x)\|^2 \tag{4}$$

the gradient can be calculated by the following formula:

$$\nabla\Psi_\mu(x)=\left[\Theta'_\mu(x)\right]^{\mathrm{T}}\Theta_\mu(x)$$

where

$$\phi'_\mu(x)=\mathrm{diag}\left(\frac{\exp\left(\frac{x_i}{\mu}\right)-\exp\left(-\frac{x_i}{\mu}\right)}{\exp\left(\frac{x_i}{\mu}\right)+\exp\left(-\frac{x_i}{\mu}\right)},i=1,2,\cdots,n\right)$$

now, we can give a neural network model for solving the absolute value equation, which is based on the steepest descent neural network model for (4).

$$\frac{\mathrm{d}x(t)}{\mathrm{d}t}=-\tau\nabla\Psi_\mu(x),\ \tau>0 \tag{5}$$

where τ is a parameter $\tau>1$ represents that one can use a larger step size in the simulation, specific details can be referred to [14] -[16] . To simplify our analysis, we let $\tau=1$ throughout this paper. A block diagram (Figure 1) of the neural network is shown as follows.

4. ANALYSIS OF STABILITY AND EXISTENCE

Next, we recall some materials about first order differential equations (ODE) [17] :

$$x'(t) = H(x(t)),\ x(t_0) = x_0 \in R \tag{6}$$

where H is a $R^n \to R^n$ mapping. We also introduce three kinds of stability that will be discussed later.

Definition 3.1 A point $x^* = x(t^*)$ is called an equilibrium point or a steady state of the dynamic system (6)

if $H(x^*)=0$ If the reisaneighborhood $\Omega^* \in R^n$ of x^* such that $H(x^*)=0$ a $H(x) \neq 0, \forall x \in \Omega^* \setminus \{x^*\}$,

then x^* is called an isolated equilibrium point.

Lemma 3.1 Assume that $H: R^n \to R^n$ is a continuous mapping. Then, for any $t_0 \geq 0$ and $x_0 \in R^n$, there exists a local solution $x(t)$ for (6) with $t \in [t_0, \tau]$ for some $\tau > t_0$. If, in addition, H is locally Lipschitz continuous at x_0, then the solution is unique.

Definition 3.2 (Asymptotic stability). An isolated equilibrium point x^* is said to be asymptotically stable if

in addition to being Lyapunov stable, it has the property that $x(t) \to x^*$ as $t \to \infty$ for all $\|x(t_0) - x^*\| < \delta$

Definition 3.3 (Lyapunov stability). Stability in the sense of Lyapunov Let $x(t)$ be a solution for (6). An isolated equilibrium point x^* is Lyapunov stable if for any $x(t_0) = x_0$ and any $\varepsilon > 0$ there exists a $\delta > 0$

such that $\|x(t)-x^*\|<\varepsilon$ for all $t\geq t_0$ and $\|x(t_0)-x^*\|<\delta$

Definition 3.4 (Lyapunov function). Let $\Omega\subseteq R^n$ be an open neighborhood of $\bar{x}$. A continuously differentiable function $W:R^n\to R^n$ is said to be a Lyapunov function at the state $\bar{x}$ over the set Ω for Equation (6) if

$$\begin{cases} W(\bar{x})=0,\ W(x)>0,\ \forall x\in\Omega\backslash\{\bar{x}\}. \\ \dfrac{\mathrm{d}W(t)}{\mathrm{d}t}=\nabla W(x(t))^{\mathrm{T}} H(x(t))\leq 0,\ \forall x\in\Omega. \end{cases}$$

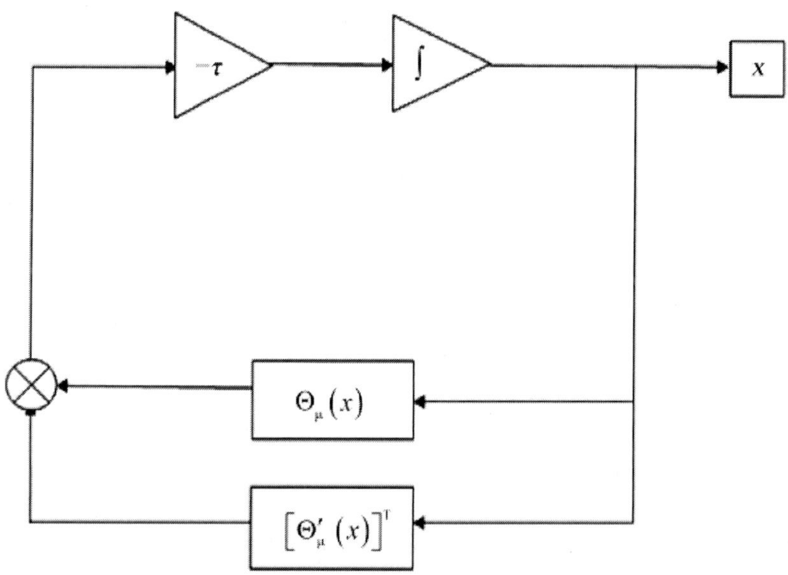

Figure 1. The block diagram of neural network (5).

Lemma 3.2 a) An isolated equilibrium point $\bar{x}$ is Lyapunov stable if there exists a Lyapunov function over some neighborhood Ω^* of x^*

b) An isolated equilibrium point $\bar{x}$ is asymptotically stable if there is a Lyapunov function over some

neighborhood Ω^* of x^* such that $\frac{dW(x(t))}{dt} < 0$ for all $x \in \Omega^* \setminus \{x^*\}$

Lemma 3.3 [11] For any $\mu > 0$, $\|A^{-1}\| < 1$, if $\nabla\Psi_\mu(x) = 0$ then $\Psi(x) = 0$.

Theorem 3.1 $\Psi_\mu(x)$ is Lyapunov function over some neighborhood Ω^* of x^*

Proof. Let x^* be the solution of the absolute value equation.

1) The function $\Psi_\mu(x)$ is obtained by our smooth approximation. So $\Psi_\mu(x)$ is continuous with respect to x. Obviously $\Psi_\mu(x)$ have continuous partial derivatives at all components of the x.

2) Since $\Theta_\mu(x) = 0$, then $\Psi_\mu(x) = \frac{1}{2}\|\Theta_\mu(x)\|^2 = 0$

3) If $x \neq x^*$ then $\Psi_\mu(x) = \frac{1}{2}\|\Theta_\mu(x)\|^2 > 0$ is always holds

So, by the Definition 3.4 we know that $\Psi_\mu(x)$ is Lyapunov function over some neighborhood Ω^* of x^*

Theorem 3.2 Each solution of the absolute value equation is the equilibrium point of the neural network (5).

Conversely, if $\|A^{-1}\| < 1$, the equilibrium point of the neural network (5) is the solution of the absolute value equation.

Proof. Assume that x^* is the solution of the absolute value equation, since $\Psi_\mu(x)=\frac{1}{2}\|\Theta_\mu(x)\|^2\geq 0$, for any $x\in R^n$ we have $\Psi(x^*)=0$ if and only if x^* is the solution of the absolute value equation.

Obviously, we got $\nabla\Psi_\mu(x^*)=0$, so x^* is the equilibrium point of the neural network (5). On the other hand if $\|A^{-1}\|<1$ and $\nabla\Psi_\mu(\bar{x})=0$, then we get $\Psi_\mu(\bar{x})=0$. So, the equilibrium point of the neural network (5) is the solution of the absolute value equation.

Next, we can prove that x^* is not only Lyapunov stable and asymptotically stable.

Theorem 3.3. Let the x^* be the isolated equilibrium of the neural network. x^* is the Lyapunov stability and asymptotic stability for neural networks.

Proof. Since x^* is the isolated equilibrium of the neural network, x^* the solution of the absolute value equation is known by the Theorem 3.2. Therefore, $\Psi_\mu(x^*)=0$. In addition, Since x is the isolated equilibrium point, so $\nabla\Psi_\mu(x^*)=0$ and $\nabla\Psi_\mu(x)\neq 0,\forall x\in\Omega^*\backslash\{x^*\}$ are hold over the neighborhood Ω^* of x^*. By

Theorem 3.1 we know that $\Psi_\mu(x)$ is Lyapunov function over some neighborhood Ω^* of x^*, so by Lemma 3.2 the isolated equilibrium x^* is Lyapunov stable. Because x^* is isolated, it is not difficult to compute:

$$\frac{\mathrm{d}\Psi_\mu(x(t))}{\mathrm{d}t}=\nabla\Psi_\mu(x(t))^{\mathrm{T}}\frac{\mathrm{d}x(t)}{\mathrm{d}t}=\nabla\Psi_\mu(x(t))^{\mathrm{T}}(-\tau\nabla\Psi_\mu(x(t)))=-\tau\|\Psi_\mu(x(t))\|\leq 0$$

Consequently, we have $\frac{d\Psi_\mu(x(t))}{dt} < 0, \forall x \in \Omega^* \setminus \{x^*\}$. By Lemma 3.2, x^* is asymptotic stability.

5. NUMERICAL EXPERIMENT

In this section we give some smooth of numerical tests of neural network algorithm, due to the complementarity problem can be transformed to absolute value equations, we consider the linear complementarity problem which is equivalent to the absolute value equations as test cases.

For a given matrix M and vector q, The linear complementarity problem $LCP(M,q)$ is to find a vector $x \in R^n$ to satisfy $x \geq 0, Mx + q \geq 0, x^{\mathrm{T}}(Mx+q) = 0$.

From the Theorem 2 in the literature [11] , if 1 is not the eigenvalues of the matrix M, then $LCP(M,q)$ is equivalent to the following absolute value equation:

$$(M+I)(M-I)^{-1}x - |x| = ((M+I)(M-I)^{-1} - I)q$$

where $x = (M-I)z + q$ and z is the solution of the absolute value equation.

Example 1 [11] . Consider the following linear complementary problem:

$$M = \begin{pmatrix} 1 & -4 & 1 & 0 \\ 0 & 1 & 0 & 1 \\ -1 & 0 & 0 & 0 \\ 0 & -1 & 0 & 0 \end{pmatrix}, \quad q = \begin{pmatrix} -5 \\ -5 \\ 1 \\ 1 \end{pmatrix}$$

Since1 is not included in the eigenvalues of M, then the linear complementary problem can be transformed into the following absolute value equation and they are equivalent: $Ax-|x|=b$

where

$$A=(M+I)(M-I)^{-1}=\begin{pmatrix}-1 & 8 & -2 & 8\\ 0 & -1 & 0 & -2\\ 2 & -8 & 1 & -8\\ 0 & 2 & 0 & 1\end{pmatrix}$$

$$b=\left((M+I)(M-I)^{-1}-I\right)q=\begin{pmatrix}-24\\ 8\\ 22\\ -10\end{pmatrix}$$

We can find that $x^*=(-1 \ \ -1 \ \ -8 \ \ -4)$ is a solution of the absolute equation.

By using the neural network model, the initial point is generated by x0 = rand (n,1), and the program is performed under the environment of MATLAB7.11.0. The following two figures (Figure 2and Figure 3) describe how the approximate solution of example 1 and the energy function varies with time.

Since $x=(M-I)z+q$, then we can capture the solution of the linear complementary problem $LCP(M,q)$, the solution is $z^*=(1 \ \ 1 \ \ 8 \ \ 4)$.

Example 2 [11] . Consider the following linear complementary problem:

$$M=\begin{pmatrix}2 & 1 & 1 & 1\\ 1 & 2 & 0 & 1\\ 1 & 0 & 1 & 2\\ -1 & -1 & -2 & 0\end{pmatrix},\quad q=\begin{pmatrix}-8\\ -6\\ -4\\ 3\end{pmatrix}$$

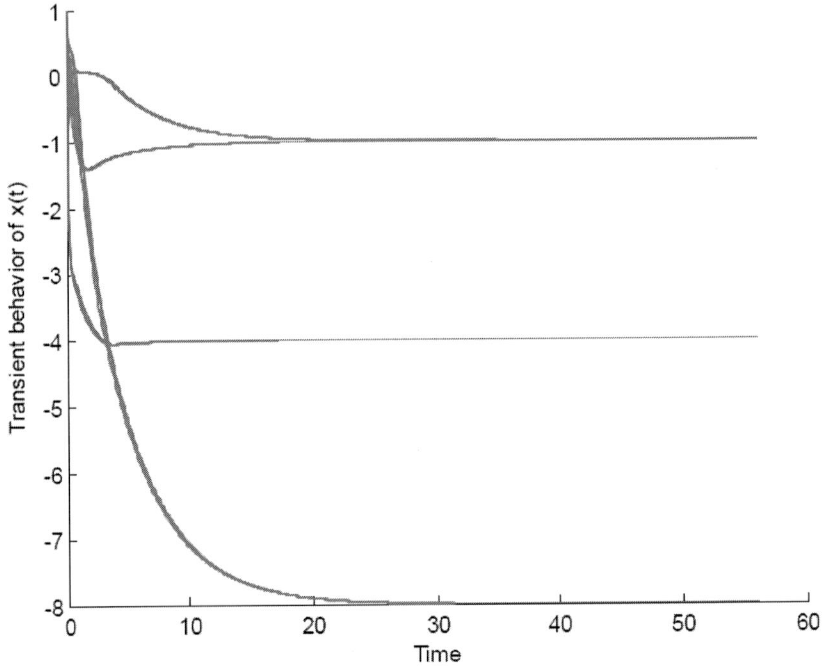

Figure 2. Transient behavior of x(t) of example 1.

Through calculation, we can get one eigenvalue of M is 1. By literature [11] , we can find that if 1 is the eigenvalue of matrix M , then the M and q of the linear complementary problem need to be multiplied by a positive constant λ and makes 1 not the eigenvalue of M (and the solution of the linear complementary pro- blem keeps invariant). Then we can transform linear complementary problem into absolute value equation by applied Theorem 2 and Theorem 3 in literature [11] .

Set $\lambda = 3$, then we can find that 1 is not included in the eigenvalues of λM . And $LCP(M,q)$ and $LCP(\lambda M, \lambda q)$ have the common optimal solution, while z can be transformed into the absolute value equation by applying the Theorem 2. Then, we have

$$M = \begin{pmatrix} 6 & 3 & 3 & 3 \\ 3 & 6 & 0 & 3 \\ 3 & 0 & 3 & 6 \\ -3 & -3 & -6 & 0 \end{pmatrix}, \; q = \begin{pmatrix} -24 \\ -18 \\ -12 \\ 9 \end{pmatrix}$$

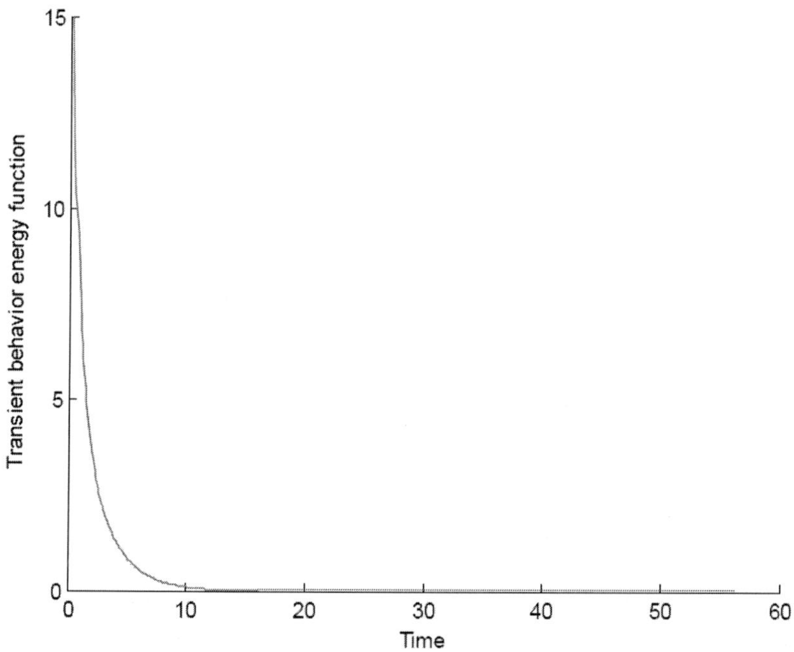

Figure 3. Transient behavior of energy function of example 1.

And the absolute value equation is $Ax-|x|=b$, where:

$$A = \begin{pmatrix} 2.0330 & -0.3626 & -0.2637 & 0.4286 \\ -0.3626 & 1.4890 & -0.0989 & -0.2143 \\ -0.2637 & -0.0989 & 1.1099 & -0.4286 \\ -0.4286 & 0.2143 & 0.4286 & 0.9286 \end{pmatrix}, \; b = \begin{pmatrix} -11.2418 \\ -0.8407 \\ 2.9341 \\ 0.6429 \end{pmatrix}$$

Thus, we can get one solution of the absolute value equation whcih is $x^{*}=(-2.5 \quad -0.5 \quad 10.5 \quad -2.5)$, then the following two figures (Figure 4 and Figure 5) describe how the approximate solution of example 2 and the energy function varies with time

Since $x=(M-I)z+q$, then we can capture the solution of the linear complementary problem LCP(M,q), the solution is $z^{*}=(2.5 \quad 2.5 \quad 0 \quad 2.5)$.

Example 3. Consider the following linear complementary problem:

$$M=\begin{pmatrix}1 & 2 & 2 & 2\\ 0 & 1 & 2 & 2\\ 0 & 0 & 1 & 2\\ 0 & 0 & 0 & 1\end{pmatrix},\ q=\begin{pmatrix}-1\\ -1\\ -1\\ -1\end{pmatrix}$$

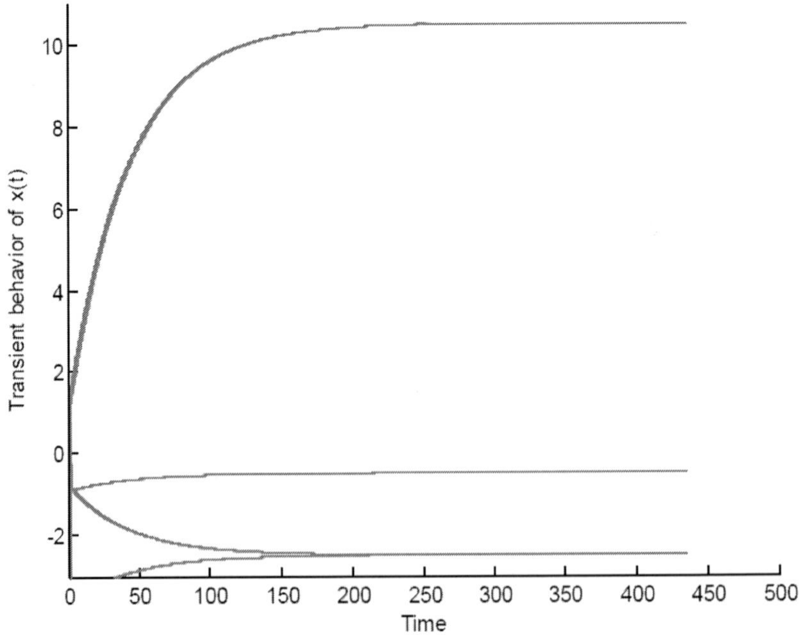

Figure 4. Transient behavior of x(t) of example 2.

Through calculation we can get one eigenvalue of M is 1. And the same as example 2, set $\lambda = 3$, then, we can find that:

$$M = \begin{pmatrix} 3 & 6 & 6 & 6 \\ 0 & 3 & 6 & 6 \\ 0 & 0 & 3 & 6 \\ 0 & 0 & 0 & 3 \end{pmatrix}, \quad q = \begin{pmatrix} -3 \\ -3 \\ -3 \\ -3 \end{pmatrix}$$

And the absolute value equation is: $Ax - |x| = b$, where:

$$A = \begin{pmatrix} 2 & -3 & 6 & -12 \\ 0 & 2 & -3 & 6 \\ 0 & 0 & 2 & -3 \\ 0 & 0 & 0 & 2 \end{pmatrix}, \quad b = \begin{pmatrix} 24 \\ -12 \\ 6 \\ -3 \end{pmatrix}$$

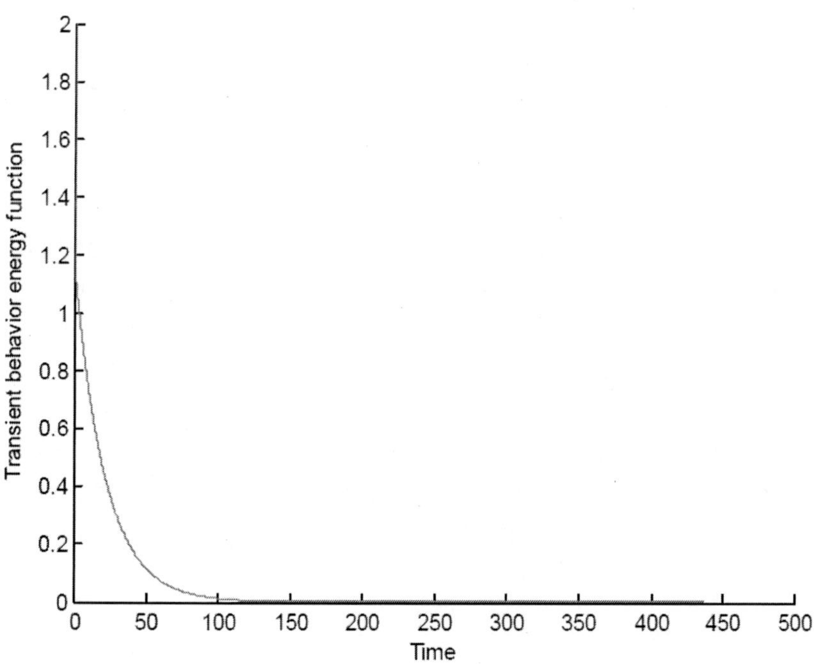

Figure 5. Transient behavior of energy function of example 2.

Thus, we can get the solution of the absolute value equation which is $x^* = (3 \quad 3 \quad 3 \quad -1)$, then the following two figures (Figure 6 and Figure 7) describe how the approximate solution of example 3 and the energy function varies with time

Since $x = (M - I)z + q$, then we can capture the solution of the linear complementary problem LCP(M,q), the solution is $z^* = (2.5 \ 2.5 \ 0 \ 2.5)$.

6. CONCLUSION

This paper adopted the aggregate function method to tackle the absolute value equation with smooth processing, and then turned the absolute value equation into a differentiable unconstrained optimization problem. In order to obtain the approximate solution of the original problem we use the proposed neural network model to solve the unconstrained optimization problem. At the same time, we propose one neural network which is based on different energy function. Through the transformation between linear complementary problem and absolute value equation, it can be used to solve the linear complementary problem, too. For the traditional energy function based on the NCP function, we can avoid a lot of matrix calculation. Numerical examples show that the algorithm is very effective for solving this kind of absolute value equation, and the accuracy of solution can be controlled by the parameters completely. In view of the fact that it is relatively difficult to solve the absolute value equation, the proposed method in this paper can be used to solve the absolute value problem effectively.

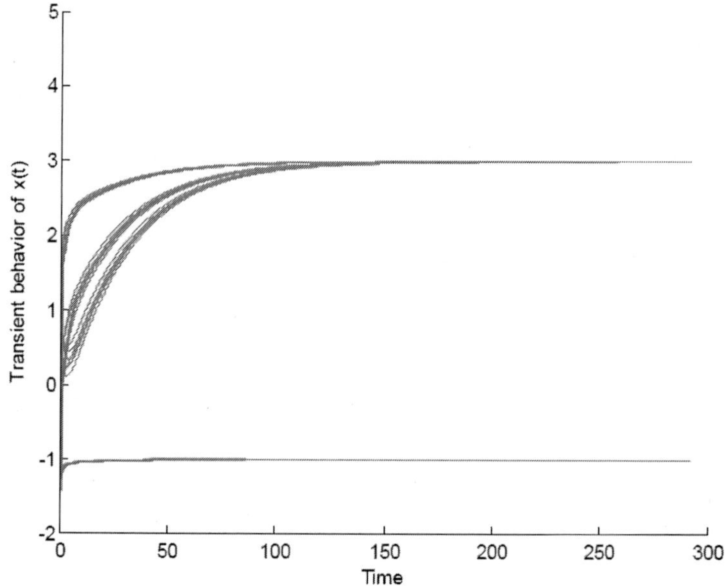

Figure 6. Transient behavior of x(t) of example 3.

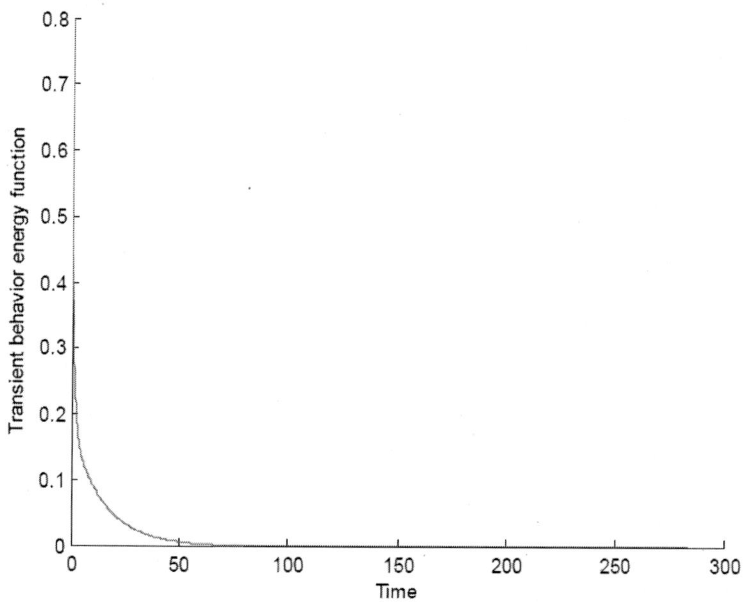

Figure 7. Transient behavior of energy function of example 3.

ACKNOWLEDGEMENTS

This work is supported by National Natural Science Foundation of China (No.11171221) and Innovation Program of Shanghai Municipal Education Commission (No.14YZ094).

REFERENCES

1. Mangasarian, O.L. and Meyer, R.R. (2006) Absolute Value Equations. Linear Algebra and Its Applications, 419, 359-367.

2. Rohn, J. (2004) A Theorem of the Alternatives for the Equation $Ax + B|x| = b$. Linear and Multilinear Algebra, 52, 421-426.

3. Han, J.Y., Xiu, N.H. and Qi, H.D. (2006) Nonlinear Complementary Theory and Algorithm. Shanghai Science and Technology Press, Shanghai.

4. Rohn, J.R. (1989) Systems of Linear Interval Equations. Linear Algebra and Its Applications, 126, 39-78.

5. Mangasarian, O.L. (2009) Knapsack Feasibility as an Absolute Value Equation Solvable by Successive Linear Programming. Optimization Letter, 3, 161-170.

6. Yamashita, S. and Fukushima, M. (2001) A Branch-and-Bound Method for Absolute Value Program and Its Application to Facility Location Problems. Kyoto University, Kyoto.

7. Mangasarian, O.L. (2013) Absolute Value Equation Solution via Dual Complementarity. Optimization Letters, 7, 625-630.

8. Mangasarian, O.L. (2009) A Generalized Newton Method for Absolute Value Equations. Optimization Letters, 3, 101-108.

9. Caccetta, L., Qu, B. and Zhou, G.L. (2011) A Globally and Quadratically Convergent Method for Absolute Value Equations. Computational Optimization and Applications, 48, 45-58.

10. Rohn, J. (2009) An Algorithm for Solving the Absolute Value Equation. Electronic Journal of Linear Algebra, 18, 589-599.

11. Yong, L.Q. and Tuo, S.H. (2012) Quasi-Newton Method to Absolute Value Equations Based on Aggregate Function. Journal of System science and Mathematics, 32, 1427-1436.

12. Yong, L.Q., Liu, S.Y. and Tuo, S.H. (2014) Transformation of the Linear Complementarity Problem and the Absolute Value Equation. Journal of Jilin University (Science Edition), 4, 638-686

13. Li, X.S. (1991) The Aggregation Function Method for Solving the Problem of Nonlinear Mini-Max Problem. Computational Structural Mechanics and Its Application, 8, 85-92.

14. Chen, J.-S., Ko, C.-H. and Pan, S.H. (2010) A Neural Network Based on the Generalized Fischer-Burmeister Function for Nonlinear Complementarity Problems. Information Sciences, 180, 697-711.

15. Ma, H.Q. and Huang, N.-J. (20145) Neural Network Smoothing Approximation Method for Stochastic Variational Inequality Problems. Journal of Industrial and Management Optimization, 7, 645-660.

16. Liao, L.Z., Qi, H.D. and Qi, L.Q. (2001) Solving Nonlinear Complementarity Problems with Neural Networks: A Reformulation Method Approach. Journal of Computational and Applied Mathematics, 131, 343-359.

17. Miller, R.K. and Michel, A.N. (1982) Ordinary Differential Equations. Academic Press, Waltham.

CHAPTER 7

An Inexact Implementation of Smoothing Homotopy Method for Semi-Supervised Support Vector Machines

Huijuan Xiong, Feng Shi

College of Science, Huazhong Agricultural University, Wuhan, China

ABSTRACT

Semi-supervised Support Vector Machines is an appealing method for using unlabeled data in classification. Smoothing homotopy method is one of feasible method for solving semi-supervised support vector machines. In this paper, an inexact implementation of the smoothing homotopy method is considered. The numerical implementation is based on a truncated smoothing technique. By the new technique, many "non-active" data can be filtered during the computation of every iteration so that the computation cost is reduced greatly. Besides this, the global convergence can make better local minima and then result in lower test errors. Final numerical results verify the efficiency of the method.

KEYWORDS

Semi-Supervised Classification; Support Vector Machines; Truncated Smoothing Technique; Global Convergence

1. INTRODUCTION

In the field of machine learning, it's essential to collect a large amounts of labeled data for the purpose of training learning algorithms. However, for many applications, huge number of data can be cheaply collected, but manual labeling of them is often a slow, expensive and error-prone process. It's desirable to utilize the unlabeled data points for the implementation of the learning task. The goal of semi-supervised classification is to employ the large collection of unlabeled data jointly with a few labeled data to finish the task of classification and prediction [11,18].

Semi-supervised support vector machines (S^3VMs) is an appealing method for the semi-supervised classification. In [7], K.P. Bennett et al. first formulated it as a mixed integer programming such that some state-ofthe-art softwares can handle the formulation. Since that, a large number of methods have been applied to solve the non-convex optimization problem associated with S^3VMs, such as convex-concave procedures [5], non-differntiable methods [1], gradient descent method [13], continuation technique [12], branch-and-bound algorithms [7,14], and semi-definite programming [17] etc. A survey of these methods can be seen from [11,18].

As pointed out in [12], one reason for the large number of proposed algorithms for S^3VMs is that the resulting optimization problem is non-convex that generates local minima. Hence, it's necessary to find better local minima because better local minima tend to lead to higher prediction accuracy. In [12], a global continuation technique is presented. In [21], a similar global smoothing homotopy method is given. However, both the method is experiential and the calculations are lengthy.

The focus of this paper is giving a new efficient implementation of the smoothing homotopy method for the S³VMs. In Section 2, we first introduce the new S³VMs model used in [21] and list two smoothing functions called aggregate function and twice aggregate function respectively. The two smoothing functions are given to approximate the nonsmooth objective function (the detailed discussion of these two smoothing functions can be seen from [4]). And then the smoothing homtopy method solving S³VMs is recalled. In Section 3, the new truncated smoothing technique is established to give a more efficient pathfollowing implementation of the smoothing homotopy method. The new technique is based on a fact that, some "non-active" data do little effect on the value of the smooth approximation functions with their gradients and Hessian during the computation, as a result, these "non-active" data can be filtered by the new truncated technique to save the computation cost. With the inexact computation technique, only a part of original data is necessary during the computation of every iteration. In the last section, Two artificial data sets with six standard test data from [10] are given to show the efficiency of our method.

A word about the notations in this paper. All vectors will be column vectors unless transposed to a row vector by a prime superscript T. The scalar (inner) product of two vectors x and y in the n-dimensional real space R^n will be denoted by $x^T y$. For a matrix $A \in R^{m\times n}$, A_i will denote the i th row of A. For a real number a, $|a|$ denotes its absolute value. For a vector $x \in R^n$, $\|x\|_1$ denotes its 1-norm, i.e., $\|x\|_1 = \sum_{i=1}^{n} |x_i|$, $\|x\|_\infty$ denotes its infty-norm, i.e., $\|x\|_\infty = \max_{1\le i\le n} |x_i|$. For an index set I, $|I|$ denotes the element number of it. For a given function $f: R^n \to R$, if f is smooth, its gradient is denoted by $\nabla f(x)$, if f is nondifferential, denote its subdifferential as $\partial f(x)$.

2. SEMI-SUPERVISED SUPPORT VECTOR MACHINES

There lies several formulations for S[3]VMs such as the mixed integer programming model by K.P. Bennett et al. [7], the nonsmooth constrained optimization model by O.L. Mangasarian [5], and the smooth nonconvex programming formulation by O. Chapelle [13] and etc. Here we mention the contributions by O. Chapelle et al. in [11-14]).

Given a dataset consisting of m labeled points and p unlabeled points all in R^n, where the m labeled points are represented by the matrix $A \in R^{m\times n}$, p unlabeled points are represented by the matrix $B \in R^{p\times n}$ and the labels for A are given by $m\times m$ diagonal matrix D of ± 1. The linear S[3]VMs to find a hyperplane $\omega^{\mathrm{T}}x = b$ far away from both the labelled and unlabeled points can be formulated as follows:

$$\begin{aligned} \min_{\omega,b} \quad & \frac{1}{2}\left(\omega^{\mathrm{T}}\omega + b^2\right) + C_1\bar{f}_1(\omega,b) + C_2\bar{f}_2(\omega,b), \\ \text{s.t.} \quad & \frac{1}{p}\left(\sum_{i=1}^{p}\omega^{\mathrm{T}}B_i + b\right) = \frac{1}{m}\sum_{i=1}^{m}D_{ii} \end{aligned} \tag{1}$$

where

$$\bar{f}_1(\omega,b) = \sum_{i=1}^{m}\max\left\{0, f_1^i(\omega,b)\right\}$$

$$\bar{f}_2(\omega,b) = \sum_{i=1}^{p}\max\left\{0, f_2^i(\omega,b)\right\}$$

$f_1^i(\omega,b)$ and $f_2^i(\omega,b)$ are loss functions corresponding to the labeled data and unlabeled data respectively and defined as follows,

$$f_1^i(\omega,b) = 1 - D_{ii}\left(\omega^{\mathrm{T}}A_i + b\right)$$

$$f_2^i(\omega,b)=1-\left|\omega^{\mathrm{T}}B_i+b\right|$$

where $\left|\omega^{\mathrm{T}}B_i+b\right|$ denotes the absolute value of $\omega^{\mathrm{T}}B_i+b$. The constraint is called balanced constraint that is used to avoid unbalanced solutions which classify all the unlabeled points in the same class.

For arbitrary vector $\xi\in R^n$, there lies an equivalent relation between its 1-norm and inf-norm in the sense that $\frac{1}{n}\|\xi\|_1\le\|\xi\|_\infty\le\|\xi\|_1$, then the sum term of model (1) can be substituted by the inf-norm form and model (1) can be reformulated as follows,

$$\begin{aligned}\min_{\omega,b}\quad & \frac{1}{2}\left(\omega^{\mathrm{T}}\omega+b^2\right)+C_1f_1^*(\omega,b)+C_2^*f_2^*(\omega,b)\\ \text{s.t.}\quad & \frac{1}{p}\left(\sum_{i=1}^{p}\omega^TB_i+b\right)=\frac{1}{m}\sum_{i=1}^{m}D_{ii}\end{aligned}\qquad(2)$$

where

$$f_1^*(\omega,b)=\max_{1\le i\le m}\left\{0,f_1^i(\omega,b)\right\}$$

$$f_2(\omega,b)=\max_{1\le i\le p}\left\{0,f_2^i(\omega,b)\right\}$$

.

We rewrite the constraint into the objective as a barrier term and reformulate $f_2^i(\omega,b)$ into its equivalent formulation $f_2^i(\omega,b)=\min\left\{1-\omega^{\mathrm{T}}B_i-b,1+\omega^{\mathrm{T}}B_i+b\right\}$, and then obtain the following formulation that is our goal in the paper.

$$\min_{\omega,b}\ \frac{1}{2}\left(\omega^T\omega+b^2\right)+C_1f_1(\omega,b)+C_2f_2(\omega,b)+f_3(\omega,b)\qquad(3)$$

where

$$f_1(\omega,b)=\max_{1\le i\le m}\left\{0,f_1^i(\omega,b)\right\}$$

$$f_2(\omega,b)=\max_{1\le i\le p}\left\{0,\min\left\{f_2^{i1}(\omega,b),f_2^{i2}(\omega,b)\right\}\right\}$$

$$f_2^{i1}(\omega,b)=1-\omega^{\mathrm{T}}B_i-b$$

$$f_2^{i2}(\omega,b)=1+\omega^{\mathrm{T}}B_i+b$$

$$f_3(\omega,b)=M\left|\frac{1}{p}\left(\sum_{i=1}^{p}\omega^{\mathrm{T}}B_i+b\right)-\frac{1}{m}\sum_{i=1}^{m}D_{ii}\right|^2$$

M is a barrier parameter.

If the dataset is nonlinear separable, we need construct a surface separation based on some kernel trick (detailed discussion of it can be seen from [2] and etc.). Denote $\bar{A}=[A;B]^{\mathrm{T}}$, assume that the surface we want to find is $K(x,\bar{A})u+b=0$, where $K(\cdot,\cdot)$ is usually taken as Gaussian kernel function with the form of

$$K(x,y)=\exp\left(-\mu\|x-y\|^2\right)$$

μ is the kernel parameter, $u\in R^{m+p}$. To find the nonlinear decision surface, we need to solve the following problem:

$$\begin{aligned}\min_{u,b}\quad f(u,b)=&\frac{1}{2}\left(u^{\mathrm{T}}u+b^2\right)+C_1f_1(u,b)\\&+C_2f_2(u,b)+f_3(u,b)\end{aligned}\tag{4}$$

where

$$f_1(u,b)=\max_{1\le i\le m}\left\{0,1-D_{ii}\left(K\left(A_i^{\mathrm{T}},\bar{A}\right)u+b\right)\right\}$$

$$f_2(u,b)$$
$$=\max_{1\le i\le p}\left\{0,1+\min\left\{K\left(B_i^{\mathrm{T}},\bar{A}\right)u+b,-K\left(B_i^{\mathrm{T}},\bar{A}\right)u-b\right\}\right\}$$

$$f_3(u,b)=M\left|\frac{1}{p}\left(\sum_{i=1}^{p}K\left(B_i^{\mathrm{T}},\bar{A}\right)u+b\right)-\frac{1}{m}\sum_{i=1}^{m}D_{ii}\right|^2$$

2.1. Aggregate Function and Aggregate Homotopy Method

Aggregate function is an attractive smooth approximate function. It has been used extensively for the the nonsmooth min-max problem [4], variational inequalities [6], mathematical programm with equilibrium constraints (MPEC) [16], non-smooth min-max-min problem [6] and etc. In the following, we will utilize the approximate function with its modification to establish an globally convergent method, called as aggregate homotopy method, for solving model (3) or (4).

In short, let $x=(\omega,b),s=n+1$ (or, $x=(u,b),s=m+p+1$) and denote model (3) or (4) as the following unified formulation:

$$\min_x f(x)=f_1(x)+f_2(x) \tag{5}$$

where

$$
\begin{aligned}
&f_1(x) = \max_{1 \le i \le m} \left\{ \bar{f}_1^i(x) \right\}, \\
&f_2(x) = \max_{1 \le i \le p} \left\{ \min \left\{ \bar{f}_2^{i1}(x), \bar{f}_2^{i2}(x) \right\} \right\}, \\
&\bar{f}_1^i(x) = \frac{1}{2} x^{\mathrm{T}} x + f_1^i(x), \\
&\bar{f}_2^{i1}(x) = f_2^{i1}(x) + f_3(x), \\
&\bar{f}_2^{i2}(x) = f_2^{i2}(x) + f_3(x),
\end{aligned} \tag{6}
$$

and $f_1^i(x), f_2^{i1}(x), f_2^{i2}(x), f_3(x)$ are defined as (3) or (4).

Denote $f_2^i(x) = \min\left\{ f_2^{i1}(x), f_2^{i2}(x) \right\}$, based on the optimality condition of non-smooth optimization theory in [9], we know that the subdifferential of $f_1(x)$ and $f_2(x)$ can be computed as follows,

$$
\begin{aligned}
&\partial f_1(x) = x + \sum_{i \in I_1(x)} \lambda_i \left[A_i^{\mathrm{T}}; 1 \right], \\
&\partial f_2(x) = \sum_{i \in I_2(x)} \eta_i \sum_{j \in J_2^i(x)} \delta_{ij} \nabla f_2^{ij}(x) + \nabla f_3(x)
\end{aligned} \tag{7}
$$

where

$$I_1(x) = \left\{ i \in \{1, \cdots, m\} : f_1^i(x) = f_1(x) \right\}$$

$$\lambda_i \in [0,1], \sum_{i \in I_1(x)} \lambda_i = 1$$

$$I_2(x) = \left\{ i \in \{1, \cdots, p\} : f_2^i(x) = f_2(x) \right\}$$

$$J_2^i(x) = \left\{ j \in \{1, 2\} : f_2^{ij}(x) = f_2^i(x) \right\}$$

$$\eta_i \in [0,1], \sum_{i \in I_2(x)} \eta_i = 1$$

$$\delta_{ij} \in [0,1], \sum_{j \in J_2^i(x)} \delta_{ij} = 1$$

$$\nabla f_2^{i1}(x) = [B_i^{\mathrm{T}}; 1]$$

$$\nabla f_2^{i2}(x) = -\nabla f_2^{i1}(x)$$

moreover, a point x^* can be called a stationary point or a solution point of (5) if satisfying $0 \in \partial f(x^*)$.

To solve model (5) by an aggregate homotopy method, we first introduce the following two smoothing functions,

$$f_1(x,t) = t \ln\left(\sum_{i=1}^{m} \exp\left(\frac{\overline{f}_1^i(x)}{t}\right)\right)$$
$$f_2(x,t) = t \ln\left(\sum_{i=1}^{p} \exp\left(\frac{\overline{f}_2^i(x,t)}{t}\right)\right) \tag{8}$$

where $\overline{f}_i^i(x)$ is defined as (6) and

$$\overline{f}_2^i(x,t) = -t \ln\left(\exp\left(-\frac{\overline{f}_2^{i1}(x)}{t}\right) + \exp\left(-\frac{\overline{f}_2^{i2}(x)}{t}\right)\right).$$

We call $f_1(x,t)$ and $f_2(x,t)$ as aggregate function and twice aggregate function respectively. The two smoothing functions have many good properties such as uniform approximation and etc. More details can be seen from [19].

Using above two uniformly approximations functions in (8), we define the following homotopy mapping:

$$H_{x^0}(x,t) = (1-t)\nabla_x f(x,t) + t(x - x^0) = 0 \tag{9}$$

where $x^0 \in R^n$ is an arbitrarily initial point and

$$\nabla_x f(x,t) = \nabla_x f_1(x,t) + \nabla_x f_2(x,t).$$

We call Equation (9) as an aggregate homotopy equation. It contains two limiting problems. On the one hand as $t=1$, it has a unique solution $x = x^0$. On the other hand, as $t \to 0$, the solution of it approaches to a stationary point of (5), i.e., a solution x satisfying $0 \in \partial f(x)$.

For a given initial point $x^0 \in R^s$, we denote the zeros point set of the aggregate homotopy mapping $H_{x^0}(\cdot,\cdot): R^s \times [0,1] \to R$ as

$$H_{x^0}^{-1}(0) = \{(x,t) \in R^s \times [0,1] : H_{x^0}(x,t) = 0\} \tag{10}$$

It can be proved that $H_{x^0}^{-1}(0)$ includes a smooth path Γ with no bifurcation points, starting from $(x^0,1)$ and approaching to the hyperplane $t=0$ that leads to a solution of the original problem [21].

3. INEXACT PREDICTOR-CORRECTOR IMPLEMENTATION OF THE AGGREGATE HOMOTOPY METHOD

The path-following of the homotopy path Γ can be implemented by some predictor-corrector procedure. Some detailed discussion on the predictor-corrector algorithm with the convergence can be seen from [3,8] and etc. In the following, we first give the framework of the predictor-corrector procedure in this paper, and then discuss how to make an inexact predictor-corrector implementation.

3.1. Predictor-Corrector Path-Following Algorithm

Parameters. initial stepsize h_0, maximum stepsize h_m, minimum stepsize h_ε, stop criteria $\varepsilon_1 > 0$ for procedure terminated, stop criteria $\varepsilon_N > 0$ for Newton corrector stopped, criteria $\varepsilon_c > 0$ for judging corrector plane, counter $N_i = 0$.

Data. $\left(x^{(0)}, t_0\right) \in R^s \times \{1\}$ Step 0. $h = h_0$, $d_0 = [0;-1]$, $k = 0$.

Step 1. Compute a predictor point $\left(\overline{x}^{(k)}, \overline{t}_k\right)$

1) Solve the following linear equation

$$\begin{bmatrix} DH\left(x^{(k)}, t_k\right) \\ d_0^T \end{bmatrix} d_1 = \begin{bmatrix} 0 \\ 1 \end{bmatrix}$$

to obtain a unit tangent vector $d_1 = \dfrac{d_1}{\|d_1\|}$;

2) $h = \min\left(h_m, h\right)$; $\left(\overline{x}^{(k)}, \overline{t}_k\right) = \left(x^{(k)}, t_k\right) + hd_1$;

3) If $\overline{t}_k < 0$ or $\overline{t}_k > 1$, $h = \dfrac{h}{2}$, return to 2); else, go to Step 2;

Step 2. Compute a corrector point $\left(x^{(k+1)}, t_{k+1}\right)$

1) If $\overline{x}_k < \varepsilon_c$, take $d = [0;1]$ and $h = \min\{h, h_\varepsilon\}$; else, take $d = d_1$. Go to 2);

2) Solve equation

$$\begin{cases} H(x,t) = 0, \\ d^T\left[x - \overline{x}^k; t - \overline{t}_k\right] = 0 \end{cases}$$

by Newton method with the stopping criteria ε_N and go to 3);

3) If Newton corrector fail, $h=0.7h$, go to 2); else, denote the solution as $\left(x^{k+1}, t_{k+1}\right)$, go to 4);

4) If $t_{k+1}>1$ or $t_{k+1}<0$, $h=\frac{h}{2}$, go to 2); else, go to 5);

5) If $t_k<\varepsilon_1$, stop ; else $d_0=d$, k = k + 1, go to 6);

6) $N_i=N_i+1$. If $N_i>3$ and the iteration number of Newton corrector is less than 3, go to 7); else , go to 1);

7) $N_i=0$, $h=1.5h$; go to 1).

Notice that, the main computation cost of Algorithm 3.1 lies in the equations solver in Step 1) and 2), some inexact computation technique can be introduce to save the computation cost of them. we take the following approximate homotopy equation $\tilde{H}(x,t)$ with its Jacobian $D\tilde{H}(x,t)$ in place of the original $H(x,t)$ with $DH(x,t)$ during the computation of step 1) and 2).

Given parameters $\varepsilon_1(x,t)>0$, $\varepsilon_2(x,t)>0$, denote $M=\{1,\cdots,m\}$, $P=\{1,\cdots,p\}$

$$I_1(x,t)=\left\{i\in M: f_1^i(x)-f_1(x)>-\varepsilon_1(x,t)\right\},$$

$$f_2^i(x,t)=-t\ln\left(\exp\left(-\frac{f_2^{i1}(x)}{t}\right)+\exp\left(-\frac{f_2^{i2}(x)}{t}\right)\right),$$

$$I_2(x,t)=\left\{i\in P: f_2^i(x,t)-\max_{i\in P}\left\{f_2^i(x,t)\right\}>-\varepsilon_2(x,t)\right\},$$

$$\tilde{f}_1(x,t)=\frac{1}{2}x^Tx+t\ln\left(\sum_{i\in I_1(x,t)}\exp\left(\frac{f_1^i(x)}{t}\right)\right),$$

$$\tilde{f}_2(x,t)=f_3(x)+t\ln\left(\sum_{i\in I_2(x,t)}\exp\left(\frac{f_2^i(x,t)}{t}\right)\right),$$

$$\begin{aligned}\tilde{H}(x,t)&=(1-t)\nabla_x\tilde{f}(x,t)+t\left(x-x^0\right)\\&=(1-t)\left(\nabla_x\tilde{f}_1(x,t)+\nabla_x\tilde{f}_2(x,t)\right)+t\left(x-x^0\right),\end{aligned}$$

$$D\tilde{H}(x,t)=\left(\frac{\partial H}{\partial x},\frac{\partial H}{\partial t}\right)$$

where

$$\frac{\partial H}{\partial x} = (1-t)\left(\nabla_x^2 \tilde{f}_1(x,t) + \nabla_x^2 \tilde{f}_2(x,t)\right) + tI_s$$

$$\frac{\partial H}{\partial t} = x - x^0 - \left(\nabla_x \tilde{f}_1(x,t) + \nabla_x \tilde{f}_2(x,t)\right) + (1-t)\left(\nabla_{xt}^2 \tilde{f}_1(x,t) + \nabla_{xt}^2 \tilde{f}_2(x,t)\right)$$

It's proven in [20] that, only if the error $\left\|\tilde{H}(x,t) - H(x,t)\right\|$ and $\left\|D\tilde{H}(x,t) - DH(x,t)\right\|$ are small enough, or equivalently, $\varepsilon_1(x,t)$ and $\varepsilon_2(x,t)$ are chosen appropriately, the approximate Euler-Newton predictor-corrector also approaches to a solution of original problem. Here we only list the main results and omit the proofs.

Denote

$$E_1(x,t) = \left\|\tilde{H}(x,t) - H(x,t)\right\|$$

and $E_2(x,t) = \left\|D\tilde{H}(x,t) - DH(x,t)\right\|$, $\tilde{d}(x,t)$ is the unit tangent vector obtained by the approximate computation, $d(x,t)$ is the tangent vector by exact Euler predictor, we have the following lemma to guarantee the efficiency of the approximate tangent vector.

3.2. Lemma

For a given $(x,t) \in \Gamma$, if $E_2(x,t)$ is small enough and satisfies

$$cond(DH)\frac{E_2(x,t)}{\|DH\|} + O\left(E_2(x,t)^2\right) < \sqrt{2}$$

The approximate unit predictor tangent vector $\tilde{d}(x,t)$ is effective, i.e., $\tilde{d}(x,t)$ still makes a direction with arclength increased.

During the correction process, the following equation must be solved

$$F(x,t)=\begin{pmatrix} H(x,t) \\ d^{\mathrm{T}}\left((x,t)-\left(x^{0},t^{0}\right)\right) \end{pmatrix}=0 \tag{11}$$

where $\left(x^{0},t^{0}\right)$ is an appropriate predictor point obtained from the former predictor step. We adopt the following approximate Newton method to solve (11),

$$\left(x^{k+1},t_{k+1}\right)=\left(x^{k},t_{k}\right)-\left(\tilde{F}\left(x^{k},t_{k}\right)\right)^{-1}\tilde{F}\left(x^{k},t_{k}\right) \tag{12}$$

where

$$\left(\tilde{F}\left(x^{k},t_{k}\right)\right)^{-1}=\begin{pmatrix} D\tilde{H}\left(x^{k},t_{k}\right) \\ d^{\mathrm{T}} \end{pmatrix}^{-1},$$

$$\tilde{F}\left(x^{k},t_{k}\right)=\begin{pmatrix} \tilde{H}\left(x^{k},t_{k}\right) \\ d^{\mathrm{T}}\left(\left(x^{k},t_{k}\right)-\left(x^{0},t_{0}\right)\right) \end{pmatrix}$$

From 0 is a regular value of $H(x,t)$ and d is a unit tangent vector induced by $\tilde{D}H(x,t)$, we know, if the step h is chosen appropriately, the equation (11) has a solution and the approximate Newton iteration (12) is effective.

3.3. Approximate Newton Corrector Convergence Theorem

Suppose that $F(x,t)=0$ have solution (x^*,t_*) with nonsingular $F'(x^*,t_*)$, there exists a neighborhood $S=\overline{S}((x^*,t_*),\delta)$ and $\gamma_1>0,\gamma_2>0$, for any $(x^0,t^0)\in S$, if for each $k=1,2,\cdots$, $E_1(x^k,t_k)$ and $E_2(x^k,t_k)$ satisfying

$$E_1(x^k,t_k)\leq\gamma_1, E_2(x^k,t_k)\leq\gamma_2\left\|F(x^k,t_k)\right\|.$$

Then the approximate Newton iteration point sequence $\{(x^k,t_k)\}$, from (12) is well defined and converges to (x^*,t_*).

4. NUMERICAL RESULTS

In this section, some numerical examples and comparisons are given to illustrate the efficiency of our method. Two artificial datasets are generated first. The first one consists of 34 points generated by “rand” function, 14 of them are labeled and the remaining 30 are seen as unlabeled data. The second one consists of 242 points taken from two nonlinear bihelix curves that are generated by $\rho=a\theta+b$, where one is obtained by taking $a=b=0.2$, the other is by taking $a=0.2,b=0.3$, $\theta\in[0:\pi/40:3\pi]$. We take randomly 30% of them as labeled and the remaining 70% as unlabeled. The comparisons of our method with the LSVM method from [15] without the consideration of unlabeled data are given. Final results are illustrated in Figures 1 and 2.

To reveal the efficiency of our algorithm for S^3VMs, comparisons of our algorithm (InSH) with some other existing algorithms for S^3VMs such as the convexconcave procedure in [5] (vS^3VM), the continuation method in [12] (cS^3VM) and the gradient descent method in [13] (∇S^3VM), are given for some standard test data from [10]. For the linear programming subproblem arising in [5], we solve it with the Matlab function *linprog* in optimization toolbox. The comparison results are listed in the following **Table 1**.

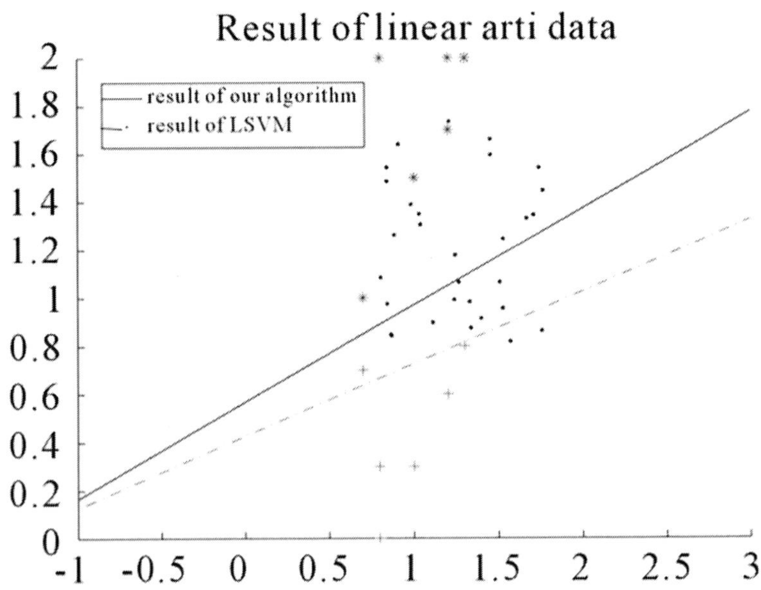

Figure 1. Result on linear artificial data of LSVM and the new algorithm.

All the computations are performed on a computer running the software Matlab 7.0 on Microsoft Windows Vista with Intel(R) Core(TM)2 Duo CPU 1.83 GHz processor and 1789 megabytes of memory. During the computation, we take $h_0 = 0.1$, $h_m = 1$, $h_\varepsilon = 1e-3$, $\varepsilon_1 = 1e-3$, $\varepsilon_N = 1e-3$, $\varepsilon_c = 1e-3$, $C, C^* \in [2^{-3}, 2^3] \cup [1e-3, 1e+2]$, $M \in [1e-3, 2]$ are taken as the one for the least test error. If necessary, the kernel parameter μ is taken the best leads to the least test error among[1e−1,3]. The parameters for determining the inexact index set are taken as

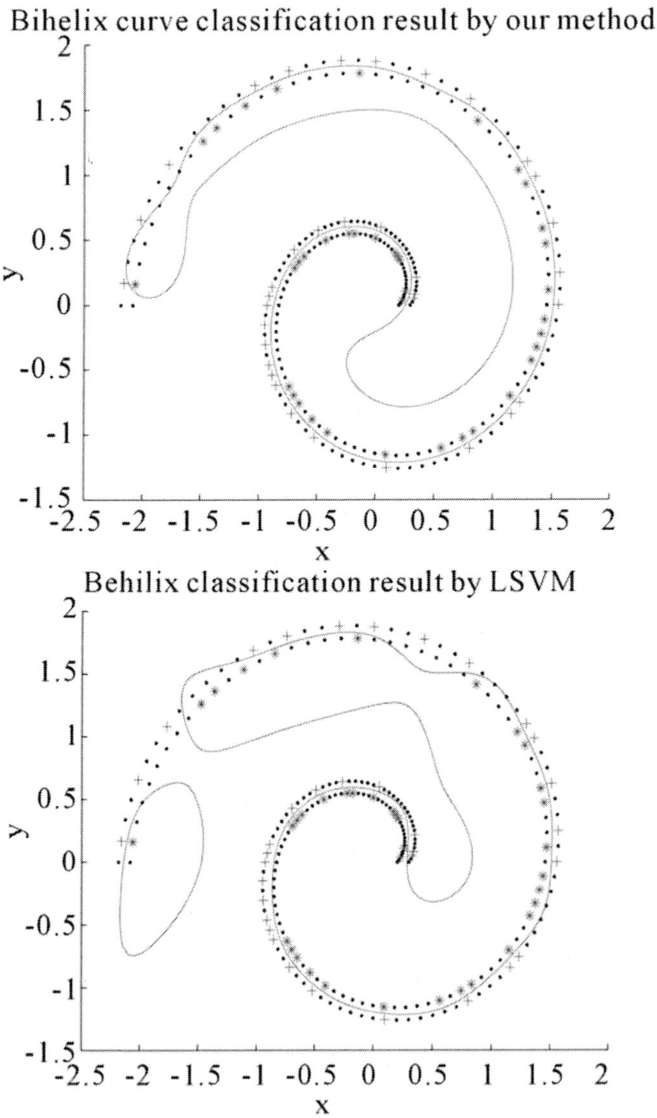

Figure 2. Result on nonlinear bihelix curve artificial data of LSVM and the new algorithm.

$$\varepsilon_1(x,t) = \begin{cases} t\ln\left(\max\left(\dfrac{4(m-1)}{\delta_2},1\right)\right), & t=1; \\ t\ln\left(2(m-1)\max\left\{\dfrac{2p}{\delta_1},\dfrac{2q}{\delta_2},1\right\}\right), & t\neq 1. \end{cases}$$

where $p = 2(1-t)B_f$, $q = 6(1-t)B_g\left(B_f/t + B_g\right)/t + 2B_g$, $B_f = \max\left(\left|f_1^i(x)\right|\right)$ and $B_g = \max\left(\max\left(|A|\right)\right)$. $\varepsilon_2(x,t)$ has the same expression as $\varepsilon_1(x,t)$ where $B_f = \max\left(\left|f_2^i(x,t)\right|\right)$ and $B_g = \max\left(\max\left(|B|\right)\right)$. δ_1 and δ_2 are taken as $\delta_1 = \delta_2 = 1\text{e}-3$ that are given to bound the error of $\|H-\tilde{H}\|$ and $\|DH-D\tilde{H}\|$.

Table 1. Comparison Results on test data (test error %).

Dataset $(m+p)\times n$	InSH (C_1, C_2, M)	vS3VM (C_1, C_2)	cS3VM (C_1, C_2)	∇S3VM (C_1, C_2)
Ionosphere (351 × 34)	11.03 (1, 0.25, 1e - 3)	11.66 (8, 2)	11.34 (100, 10)	13.37 (1, 0.1)
MUSK1 (476 × 166)	12.26 (1, 1, 0.1)	15.93 (2, 2)	15.75 (1, 1)	15.34 (100, 10)
NDC (1000 × 32)	13.60 (2, 2, 0.05)	13.90 (4, 2)	14.60 (100, 1)	24.00 (0.01, 0.1)
Pima (769 × 8)	22.66 (8, 2, 0.05)	22.65 (10, 1)	23.42 (8, 0.5)	39.45 (10, 1)
Sonar (208 × 60)	15.45 (16, 0.5, 0.01)	23.02 (100, 1)	23.10 (100, 10)	23.50 (100, 10)
Votes (435 × 16)	3.44 (8, 2, 2)	4.59 (1, 0.1)	-	4.38 (10, 1)
Aver. Time (sec.)	5.7367	9.2271	7.3925	6.0676

-: denotes the method fails for the dataset.

ACKNOWLEDGEMENTS

The research was supported by the National Nature Science Foundation of China (No. 11001092) and the TianYuan Special Funds of the National Natural Science Foundation of China (Grant No. 11226304).

REFERENCES

1. A. Astorino and A. Fuduli, "Nonsmooth Optimization Techniques for Semi-Supervised Classification," IEEE Transactions on Pattern Analysis and Machine Intelligence, Vol. 29, No. 12, 2007, pp. 2135-2142.

2. C. J. C. Burges, "A Tutorial on Support Vector Machines for Pattern Recognition, "Data Mining and Knowledge Discovery, Vol. 2, No. 2, 1998, pp. 121-167.

3. E. L. Allgower and K. Georg, "Numerical Continuation Methods: An Introduction," Springer-Vergal, Berlin, 1990.

4. E. Polak, J. O. Royset and R. S. Womersley, "Algorithms with Adaptive Smoothing for Finite Minimax Problems," Journal of Optimization Theory and Application, Vol. 119, No. 3, 2003, pp. 459-484.

5. G. Fung and O. Mangasarian, "Semi-Supervised Support Vector Machines for Unlabeled Data Classification," Optimization Methods and Software, Vol. 15, No. 1, 2001, pp. 29-44.

6. G. X., Liu, "Aggregate Homotopy Methods for Solving Sequential Max-Min Problems, Complementarity Problems and Variational Inequalities," PhD Thesis, Jilin University, Jilin, 2003.

7. K. Bennett and A. Demiriz, "Semi-Supervised Support Vector Machines," In: M. S. Kearns, S. A. Solla and D. A. Cohn, Eds, Advances in Neural Information Processing Systems, MIT Press, Vol. 10, 1998, pp. 368-374.

8. L. T. Watson, S. C. Billups and A. P. Morgan, "Algorithm 652 Hompack: A Suite of Codes for Globally Convergent Homotopy Algorithms," ACM Transactions on Mathematical Software, Vol. 13, No. 3, 1987, pp. 281- 310.

9. M. M. Mkela and P. Neittaanmki, "Nonsmooth Optimizatin: Analysis and Algorithms with Application to Optimal Control," Utopia Press, Singapore, 1992.

10. P. M. Murphy and D. W. Aha, "UCI Repository of Machine Learning Databases. http://www.ics.uci.edu/ mlearn/ML Repository.html.

11. O. Chapelle, V. Sindhwani and S. S. Keerthi, "Optimization Techniques for Semi-Supervised Support Vector Machines," Journal of Machine Learning Research, Vol. 9, 2008, pp. 203-233.

12. O. Chapelle, M. Chi and A. Zien, "A Continuation Method for Semi-Supervised SVMs," ACM International Conference Proceeding Series, Proceedings of the 23rd international conference on Machine learning, Vol. 148, 2006, pp. 185-192.

13. O. Chapelle and A. Zien, "Semi-Supervised Classification by Low Density Separation," Proceedings of the Tenth International Workshop on Artificial Intelligence and Statistics, Vol. 1, 2005, pp. 57-64.

14. O. Chapelle, V. Sindwani and S. Keerthi, "Branch and Bound for Semi-Supervised Support Vector Machines," Advances in Neural Information Processing Systems 19, Proceedings of the 2006 Conference, MIT Press, Cambridge, 2007, pp. 217-224.

15. O. L. Mangasarian and D. R. Musicant, "Lagrangian Support Vector Machines," Journal of Machine Learning Research, Vol. 1, 2001, pp. 161-177.

16. S. Birbil, S. C. Fang and J. Han, "Entropic Regularization Approach for Mathematical Programs with Equilibrium Constraints," Technical Report, Industrial Engineering and Operations Research, Carolina, 2002.

17. T. D. Bie, N. Cristianini, "Semi-Supervised Learning Using Semi-Definite Programming," In: O. Chapelle, B. Scholkopf and A. Zien, Eds., Semi-Supervised Learning, MIT Press, Cambridge, 2006.

18. X. J. Zhu, "Semi-Supervised Learning Literature Survey," Technical Report 1530, Computer Science, University of Wisconsin-Madison, 2005.

19. X. S. Li and S. C. Fang, "On the Entropic Regularization Method for Solving Min-Max Problems with Applications," Mathematical Methods and Operations Research, Vol. 46, No. 1, 1997, pp. 119-130.

20. Y. Xiao, H. J. Xiong and B. Yu, "Truncated Aggregate Homotopy Method for Nonconvex Nonlinear Programming," Optimization Methods and Software, 2012, pp. 1- 18.

21. H. J. Xiong and B. Yu, "Aggregate Homotopy Method for Semi-Supervised SVMs," 2011 International Conference on Electric Information and Control Engineering, pp. 1147-1150.

CHAPTER 8

A Robust Kalman Framework with Resampling and Optimal Smoothing

Thomas Kautz * and Bjoern M. Eskofier

Digital Sports Group, Pattern Recognition Lab, Friedrich-Alexander University Erlangen-Nürnberg (FAU), Haberstr. 2, 91058 Erlangen, Germany

ABSTRACT

The Kalman filter (KF) is an extremely powerful and versatile tool for signal processing that has been applied extensively in various fields. We introduce a novel Kalman-based analysis procedure that encompasses robustness towards outliers, Kalman smoothing and real-time conversion from non-uniformly sampled inputs to a constant output rate. These features have been mostly treated independently, so that not all of their benefits could be exploited at the same time. Here, we present a coherent analysis procedure that combines the aforementioned features and their benefits. To facilitate utilization of the proposed methodology and to ensure optimal performance, we also introduce a procedure to calculate all necessary parameters. Thereby, we substantially expand the versatility of one of the most widely-used filtering approaches, taking full advantage of its most prevalent extensions. The applicability and superior performance of the proposed methods are demonstrated using simulated and real data. The possible areas of applications for the presented analysis procedure range from movement analysis over medical imaging, brain-computer interfaces to robot navigation or meteorological studies.

KEYWORDS

Kalman filter; fixed-lag smoothing; outlier detection; real-time filtering; non-uniform sampling; parameter estimation

1. INTRODUCTION

The Kalman filter [1] is the optimal filter for linear Gaussian systems in a least mean square sense [2]. However, when the necessary preconditions, *i.e.*, the linearity of the system and Gaussian distribution of the noise terms, are violated, the optimality of the filter output cannot be guaranteed anymore.

To overcome these limitations and to enhance the performance of the KF, numerous modifications have been proposed. The adaption of the Kalman framework to non-linear systems [3], Kalman smoothing [4] and the consideration of outliers [5] are amongst the most popular of these alterations. Usually, these modifications are treated separately, and combinations of several methods are disregarded. Additionally, although the KF can be used for non-uniformly sampled data in general, most of the published work describing the aforementioned modifications assume observations that are acquired with a constant sampling rate. Finally, the applicability of the published methods for the estimation of the crucial filter parameters is tightly restricted with regard to the required preconditions. A filter framework that includes the efficient and robust handling of non-linear systems, outliers and non-uniform sampling, while at the same time allowing Kalman smoothing and automated parameter estimation, would make it possible to include all of these modifications and their advantages together. To overcome the irreconcilability of the aforementioned features, we introduce an innovative filtering framework in this paper and a method to estimate the required parameters. An overview of the five presented features and their combination is depicted in Figure 1.

Figure 1. Overview of the presented features and their combination with the original KF.

The first and arguably the most prevalent modification of the KF is its adaption to non-linear systems. In contrast to the linear KF, the extended Kalman filter (EKF) [6], the unscented Kalman filter (UKF) [7] and the square-root unscented Kalman filter (SRUKF) [8] are also applicable to non-linear systems. In the EKF, this is achieved via point-wise linearization of the process-model, whereas the UKF and SRUKF both employ the unscented transformation of the state covariance matrix. Non-linearities can also be handled with particle filters, which allow a non-parametric description of probability distributions in the state-space by Monte Carlo approximations [9]. We focus our work on the EKF, due to its high practical relevance.

Secondly, different strategies have been proposed to mitigate the effect of outliers, since they violate the assumption of Gaussian noise. According to Grubbs, “an outlying observation, or outlier, is one that appears to deviate markedly from other members of the sample in which it occurs" [10]. Outliers can occur, for example, in a radar-based localization scenario, where the measurements are disturbed by transient changes from line-of-sight to non-line-of-sight conditions. In an approach presented by Agamennoni *et al.* [11], outliers are taken into account by modeling measurement noise with a Student's t distribution instead of a Gaussian distribution. In the update step of the KF, the measurement noise covariance matrix R is replaced by its expected value, which is computed iteratively for each update. This leads to computational overhead, especially when Kalman smoothing needs to be performed, since this iterative algorithm needs to be applied several times for every output value. Moreover, the transfer of this technique to the EKF is not mentioned. A weighted robust KF is presented by Ting *et al.* [12]. Here, a scalar weight is assigned for each data sample, such that a measurement with a lower weight has a smaller contribution for the estimation of the current state. This approach performed as well as a thresholded KF, where measurements are rejected as outliers if their Mahalanobis distance exceeds a certain value. Gandhi and Mili [13] use redundant observations and a threshold on a median-based distance criterion for detection of outliers, at the cost of considerably increased computation time. Other thresholded KFs with the rejection of outliers were reported by Meyr and Spies [14] and by Mirza [15]. However, if information about the statistical properties of the outliers is available, the hard rejection of outliers might not be the optimal solution [16]. A possible alternative is to account for outliers by means of Huberization in a modified correction step. This method is described by Ruckdeschel [17], who presented two robust versions of the KF that are applicable in the presence of endogenous or exogenous outliers. Based on these filters, an outlier-robust Kalman smoother is introduced in [18]. A broader overview about robustness in the state-space context can, for example, be found in the articles by Stockinger and Dutter [19], Ershov and Lipster [20] or Martin and Raftery [21].

Most publications dealing with KFs implicitly or explicitly assume a constant sampling rate for the observations, including multirate systems that comprise several measurement units with a constant sampling rate for each of them [22,23]. Li *et al.* [24] described Kalman filtering for non-uniformly-sampled multirate systems, but outliers and smoothing were not considered. Various possibilities for delayed smoothing of the data have been presented to exploit all available data within a predefined delay, if no real time processing is required. This includes different forms of fixed-point, fixed-interval and fixed-lag smoothing [25]. Fixed-point smoothing can be used when searching for a state estimate at a fixed time, whereas fixed-interval smoothing yields state estimates for a fixed interval of measurements. In many applications (e.g., medical imaging [26], atmospheric process studies [27], target tracking [28]), where the delay that is introduced by smoothing needs to be constant, fixed-lag filtering is the most suitable choice. Considering its wide range of applications, our work focuses on fixed-lag smoothing.

Regardless of the employed modifications of the KF, the estimation of the measurement and process noise covariance matrices is essential for the optimal performance of a KF. This task is considered in various publications. An estimation scheme for a KF based on statistical properties of the innovation sequence is presented by Wang *et al.* [29]. Ghahramani and Hinton [30], as well as Axelsson and Orguner [31] proposed expectation maximization (EM) [32,33] to determine the noise parameters. Furthermore, Bavdekar *et al.* [34] introduced two different approaches (direct optimization and EM) for noise estimation. None of these publications considered the effect of outliers on the parameter estimation. In [11], outliers were taken into account, but here, a constant sampling rate and a linear model were assumed. Moreover, the utilized EM algorithm for parameter estimation was not clarified by the authors.

Despite the vast amount of published work in the field of Kalman filtering, no approach has been published so far for non-linear, outlier-robust Kalman filtering of non-uniformly sampled data with automatic estimation of all necessary filter parameters. Therefore, with the current state-of-the-art, it is not possible to make optimal use of the available data. For example, for non-uniformly-sampled data, one needs to forgo the advantages of smoothing or accept that no fixed delay can be guaranteed.

A filtering framework encompassing the five aforementioned features could improve and facilitate signal filtering in a wide variety of problems, ranging from robot navigation over movement analysis to medical imaging or TV graphics. The purpose of this paper is to introduce an innovative KF framework that combines non-linear systems, outlier handling, non-uniform sampling, fixed-lag smoothing and parameter estimation in a coherent analysis procedure.

First, the basic EKF algorithm will be outlined, and a method for the rejection of outliers will be introduced. Subsequently an EKF-based scheme to create filtered output with a constant rate from non-uniformly-sampled input will be described. Hereafter, these approaches will be combined with fixed-lag smoothing. Finally, a method for the estimation of all filter parameters under the stated conditions will be presented. The efficacy of the developed filter will then be demonstrated and discussed using real and artificial data.

2. METHODS

2.1. The Extended Kalman Filter

The original KF was designed under the assumption of linearity. With the EKF, this approach is transferred to non-linear problems by means of linearization. Like the linear KF, the EKF is a recursive filter comprising a prediction step and a measurement update step.

In the prediction step of each discrete time point k, the *a priori* state estimate of point $\hat{x}_{k|k-i}$ is calculated based on information available up to point $k-1$. In the update step, the *a posteriori* state estimate $\hat{x}_{k|k}$ is determined based on information available up to point k. The corresponding covariance matrices describing the uncertainty of the state estimates are denoted by $P_{k|k-1}$ and $P_{k|k}$. The function $F(\hat{x}_{k|k}, u_k, T_k)$ characterizes the transition of the state from time step k to time step $k+1$ and specifies the process model. The control input is denoted by u_k and T_k is the possibly non-uniform time between step $k-1$ and step k. The prediction step can be described by:

$$\hat{x}_{k+1|k} = F(\hat{x}_{k|k}, u_k, T_k) \tag{1}$$

$$P_{k+1|k} = \Phi_k P_{k|k} \Phi_k^T + \hat{\Gamma}_k Q \hat{\Gamma}_k^T \tag{2}$$

Here, Φ_k is the Jacobian of the process model for step k:

$$\Phi_k = \left[\frac{\delta F(x, u, T)}{\delta x} \right]_{\hat{x}_{k|k}, u_k, T_k} \tag{3}$$

The matrix Q represents the zero-mean Gaussian process noise. It is transformed using Γ_k, which expresses the effect of unknown disturbances on the state dynamics and depends on the choice of disturbance representation [34].

The sensor model, which relates the state vector to a measurement y_k, is represented by the function $h(\hat{x}_{k|k-1})$ with the Jacobian:

$$C_k = \left[\frac{\delta h}{\delta x} \right]_{\hat{x}_{k|k-1}} \tag{4}$$

Based on the *a priori* estimate and the measurement at step k, the a *posteriori* state estimate is calculated in the update step given by:

$$e_k = y_k - h(\hat{x}_{k|k-1}) \tag{5}$$

$$\Sigma_k = C_k P_{k|k-1} C_k^T + R \tag{6}$$

$$K_k = P_{k|k-1} C_k^T \Sigma_k^{-1} \tag{7}$$

$$x_{k|k} = x_{k|k-1} + K_k e_k \tag{8}$$

$$P_{k|k} = [I - K_k C_k] P_{k|k-1} \tag{9}$$

The term e_k is called the innovation, and Σ_k denotes the corresponding covariance matrix. R represents the measurement noise covariance matrix; K_k is the Kalman gain matrix; and I denotes the identity matrix.

2.2. Handling Outliers

For Kalman filtering, a Gaussian distribution of the measurement noise is assumed. However, there are cases where this nominal condition is violated by the presence of outliers. It is therefore necessary to make the KF robust against outliers, *i.e.*, it should have good performance under nominal conditions, but also acceptable performance, when these conditions are not met [35]. Outliers corrupt the functionality of the EKF and may even lead to divergence. To avoid such problems, the filter framework must be modified to deal with outliers. We consider the case of Type I outliers [36], *i.e.*, it is assumed that outliers are introduced into the filtering process in the form of outlying observations and are therefore non-propagating. Although there is no general quantitative definition for outliers, a widely-used specification is that from Tukey [37]. Here, extreme outliers are those data points that are not within the interval $[q_1 - 1.5 \cdot (q_3 - q_1); q_3 + 1.5 \cdot (q_3 - q_1)]$, where q_1 and $q_3$are the first and third quartiles. For normal distributions, this interval translates to approximately $\mu \pm 2.7\ \sigma$, where μ and σ are the mean and standard deviation of the underlying distribution.

One possibility to make an EKF more robust is the rejection of outliers based on a threshold-based detection: when an observation is marked as an outlier, it is not included in the filtering process. To detect outliers, we propose to evaluate the Mahalanobis distance (d_k) of the current innovation:

$$d_k = e_k^T \Sigma_k^{-1} e_k \tag{10}$$

If the Mahalanobis distance of an observation exceeds the predefined threshold α, the current update step is omitted, and the respective measurement is marked as invalid.

In theory, the outlier definition from [37] can be transferred directly into a threshold for the Mahalanobis distance, *i.e.*, $\alpha \triangleq 2.7$. However, in practice, the choice of an appropriate threshold is dependent on the adequacy of the state prediction model and the correct choice of Q and R, since they influence Σ_k. If the underlying covariance matrix is erroneous, the outliers cannot be detected reliably from the resulting Mahalanobis distance or valid measurements might wrongly be discarded. Therefore, the choice of a pertinent threshold goes along with the estimation of Q and R, which will be described in Section 2.5.

2.3. Conversion of Non-Uniform Sampling

In many applications, an estimate about the state of a system is required at predefined times or with a certain rate. This is the case when observations from different measurement systems with different sampling characteristics need to be synchronized, when transmission standards have to be met or when a constant sampling rate is necessary for further signal processing.

If no observations are available for the predefined instances, a sample rate conversion scheme needs to be employed to guarantee the availability of the necessary information when it is required. To this end, we propose to exploit the prediction abilities of the EKF.

When an observation is usable, the basic recursive steps of prediction and update are applied. State estimates for points in time at which no observation exists are computed from the previous *a posteriori* state estimate, as described in Equations (1)and (2), without a subsequent update step.

We define M as the set of time indices k for which an observation is available. The set O contains all k for which a filtered output needs to be computed (M and O need not be disjoint). The set M_v contains all k for which a valid measurement is available, *i.e.*, where the measurements were not marked as outliers ($M_v \subseteq M$).

The sample rate conversion scheme is summarized in Figure 2. The observation times are assumed to be known, but irregularly spaced. Most commercially available measurement systems provide a timestamp for each observation if the sampling rate is not constant. In the case of uniform sampling, knowledge about the sampling rate is sufficient. The points in time for which an output needs to be created can be chosen either at a constant interval or irregularly spaced. Thus, a predefined output rate can be achieved even when the input is sampled non-uniformly.

The insertion of intermediate prediction steps as described above mitigates a well-known problem: The linearization that is part of the EKF can lead to unstable filter behavior if the time step intervals are not small enough [38]. By adding additional steps, the linearization errors of the EKF can be minimized, since the step width for which the state covariance matrix needs to be propagated is reduced [39].

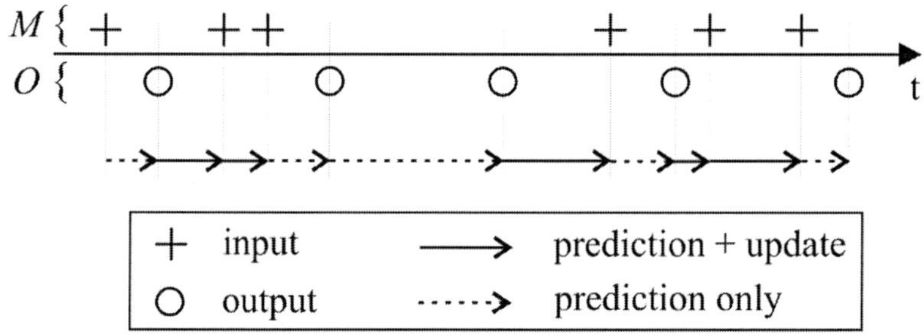

Figure 2. The sample rate conversion scheme uses the prediction abilities of the EKF. At times t for which an observation (+) is available, prediction and update are performed. When an output (o) needs to be created without a measurement, only the prediction step is conducted, and the update step is omitted. The set M contains the discrete time points for which an observation is available, and the set O contains those for which a filtered output is calculated.

2.4. Fixed-Lag Smoothing for Variable Sampling Rates

For offline data processing, the accuracy of the filtering results can be improved by employing optimal smoothing [25]. An efficient algorithm for optimal smoothing is the Rauch-Tung-Striebel (RTS) smoother [40]. For many applications, offline processing with RTS smoothing is not acceptable, and the filter output needs to be calculated with a constant delay. This can be accomplished using fixed-lag Kalman smoothing. However, to the best knowledge of the authors, no method for fixed-lag smoothing of measurements with variable sampling rates has been reported yet.

For basic RTS smoothing, the smoothed state estimates $\hat{x}_{k|N}$ and the corresponding covariance matrices $P_{k|N}$ can be calculated as follows:

$$J_k = P_{k|k}\Phi_k^T P_{k+1|k}^{-1} \tag{11}$$

$$\hat{x}_{k|N} = \hat{x}_{k|k} + J_k(\hat{x}_{k+1|N} - \hat{x}_{k+1|k}) \tag{12}$$

$$P_{k|N} = P_{k|k} + J_k(P_{k+1|N} - P_{k+1|k})J_k^T \tag{13}$$

where N is the index of the point in time corresponding to the last accessible observation and $k = N - 1, \ldots, 0$. This algorithm is valid both for uniformly- and non-uniformly-sampled data. It is worth noting that for points in time where no observation is available ($k \notin M_v$), no measurement update is possible, and $\hat{x}_{k|k}$ and $P_{k|k}$ cannot be computed.

For a fixed-lag smoother, Equations (11)–(13) need to be modified, since the data need to be processed with a constant delay L, which means that not all *a posteriori* state estimates up to time step N are available. Instead, for each k, only data up to time step l_k can be used, where $t(l_k) \leq t(k) + L$. Therefore, for fixed-lag smoothing, we propose to change Equations (12)and (13) to:

$$\hat{x}_{k|l_k} = \hat{x}_{i|i} + J_i(\hat{x}_{i+1|l_k} - \hat{x}_{i+1|i}) \tag{14}$$

$$P_{k|l_k} = P_{i|i} + J_i(P_{i+1|l_k} - P_{i+1|i})J_i^T \tag{15}$$

for $k = 1, \ldots, N$ and $i = l_k - 1, \ldots, k$ for each k. The initial point is given by $\hat{x}_{lk|lk}$ and $P_{lk|lk}$.

Again, these equations are only valid for points in time where a valid observation is available. If smoothed state estimates need to be computed at points i for which no valid measurements are available (e.g., if a constant output rate should be created from a variable input rate and/or if observations were discarded as outliers), the update step is inapplicable, and it follows that:

$$\hat{x}_{i|i} = \hat{x}_{i|i-1} \tag{16}$$

and

$$P_{i|i} = P_{i|i-1} \ \forall i \notin M_v \tag{17}$$

Thus, the *a posteriori* state features $\hat{x}_{i|i}$ and $P_{i|i}$ in Equations (14) and (15) can be replaced by the corresponding *a priori*state estimate and covariance $\hat{x}_{i|i-1}$ and $P_{i|i-1}$.

To decrease computational overhead, it is sufficient to iterate through $i \in \{l_k - 1, \ldots, k \cap (M_v \cap k)\}$, because $\hat{x}_{i|i-1}$ and $\hat{x}_{i+1|i}$both incorporate the information from the same observations. Additionally, since an output is not required for all k, smoothed estimates only need to be calculated for $k \in O$.

Depending on the available measurements and the defined delay, there may be cases where, for a given output time, there are no valid observations within the delay To ensure continuity in the filtered output, we propose to calculate the corresponding state using the usual EKF equations (including the rejection of outliers) starting from the previous filter output. The filtering scheme described above is visualized in Figure 3.

2.5. Parameter Estimation

The performance of an EKF depends crucially on the correct choice of the filter parameters. To avoid inconvenient and possibly imprecise manual tuning of the parameters, an automated estimation approach is indispensable.

We propose a new method, based on the direct optimization strategy presented in [34], to automatically estimate the required filtering parameters, *i.e.*, the process noise covariance matrix Q, the measurement noise covariance matrix R and the threshold α for the rejection of outliers. Our method does not require reference measurements for parameter estimation, since they are usually difficult to obtain or completely unavailable in practical applications. The problem of finding estimates $\hat{Q}$ and $\hat{R}$ is solved by formulating it as an optimization problem. The objective function, which needs to be minimized, is the negative of the log-likelihood function of the innovations, which can be formulated as:

$$(\hat{Q},\hat{R}) = arg \min_{(Q,R)} \left[\sum_{k=1}^{N} log(det(\Sigma_k)) + e_k^T \Sigma_k^{-1} e_k \right] \quad (18)$$

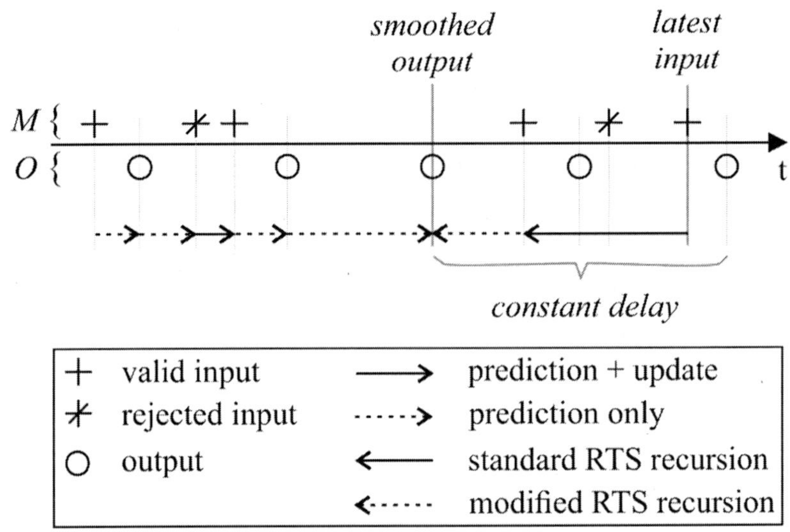

Figure 3. The sample rate conversion as shown in Figure 2 can be combined with outlier rejection and fixed-lag smoothing. Measurements that are marked as outliers are not included in the forward state estimation by skipping the update step. To guarantee a constant delay for fixed-lag smoothing, the recursion is started from the latest valid measurement and is continued up to and including the point in time for which the smoothed output is calculated. The last recursion step is conducted using the modifications described in Equations (16) and (17). Observations that were identified as outliers are ignored in the recursion procedure.

This maximum likelihood approach is also applicable for irregularly-sampled systems. It may be noted that the last part ofEquation (18) $(e_k^T \Sigma_k^{-1} e_k)$ represents the Mahalanobis distance d_k of the innovation. The term $log(det(\Sigma_k)) + e_k^T \Sigma_k^{-1} e_k$ will be summarized as f_k. In the aforementioned publication, it was proposed to solve this problem using standard nonlinear optimization methods, like sequential quadratic programming (SQP). SQP is a gradient-based optimization method and may converge to local minima without reaching the global minimum. Also the EM algorithm for parameter estimation is only guaranteed to converge to a local extremum. To avoid this, we propose to employ a global search method, such as a genetic algorithm (GA) [41], instead. Reasonable bounds for the GA can be identified based on physical considerations of the underlying process and measurement hardware.

To apply the proposed method to the outlier-robust EKF with sample rate conversion, we propose some further alterations: First, since the estimation of Q and R is based solely on the observations, all steps $k \notin O \backslash M$ can be omitted. Second, the described approach is only valid for Gaussian noise. Outliers violate the assumption of Gaussianity and, thus, distort the estimation of the true noise parameters by introducing implausibly high values of d_k in the objective function. This can lead to non-optimal estimates of the noise parameters. Additionally, an appropriate threshold needs to be found for the detection and rejection of outliers, which, in turn, depends on the correct estimates of Q and R (see Section 2.2). This interdependence calls for an approach that allows the simultaneous estimation of the noise parameters and the detection threshold α. Thus, the parameter set that needs to be estimated consists of $\Theta' = (\hat{Q}, \hat{R}, \alpha')$. The necessary modifications of the objective function to estimate not only the noise covariance matrices, but also an estimate for the optimal threshold for the detection of outliers are shown in the following.

The optimal outlier detection threshold allows the rejection of outliers and, at the same time, does not lead to the dismissal of valid measurements. To include this goal in the objective function, we propose to limit the Mahalanobis term inEquation (18) by introducing a penalty term β. If the Mahalanobis distance of the measurement at step k is smaller than the detection threshold α, the measurement is not regarded as an outlier, and f_k is calculated as stated before. If the measurement is rejected, because the Mahalanobis distance exceeds the detection threshold, d_k is replaced by β, which is defined as:

$$\beta(\Theta) = \tilde{d} + 1.5 \cdot (q_{3,d} - q_{1,d}) \tag{19}$$

where d, $q_{1,d}$ and $q_{3,d}$ are the median, lower and upper quartiles of the Mahalanobis distances d. This means that the penalty term corresponds to the boundary between outliers and inliers in the Mahalanobis distances according to the definition from [37]. As a result, the rejection of valid measurements is penalized, since, in this case, d_k would be smaller than β, *i.e.*, dismissing a valid measurement would lead to an increase of the objective function. The rejection of outliers leads to a decrease in the objective function, because here, d_k would be bigger than β.

The proposed problem formulation can be summarized as:

$$\hat{\Theta} = arg \min_{\Theta} \left[\sum_{k \in M}^{N_M} log(det(\Sigma_k)) + g_k \right] \tag{20}$$

where:

$$g_k = \begin{cases} e_k^T \Sigma_k^{-1} e_k & d_k \leq \alpha \\ \beta & d_k > \alpha \end{cases} \tag{21}$$

and N_M is the number of measurements. The genetic algorithm for the minimization of the objective function is initialized with a constrained random population of possible parameter sets. To evaluate the objective function, the EKF is tested without smoothing with each of these parameter sets. Individuals (parameter sets) that lead to high values in the objective function are discarded. Individuals that lead to low values in the objective function are propagated and modified (using crossover and mutation) by the algorithm to form a new population of parameter sets. This process is repeated, until the improvement in the objective function falls below a threshold. Further details about optimization by means of genetic algorithms can be found in [42].

Assuming the noise characteristics to be constant, parameters can be determined offline using a representative training dataset and then be applied to new data for real-time or fixed-lag filtering. If the system dynamics or the measurement quality vary over time, the proposed method can be used to estimate different parameter sets for different system states and then change between these estimates during runtime. Alternatively, other adaptive methods, as described in [43], can be combined with the proposed filter framework.

3. EXPERIMENTS AND RESULTS

3.1. Simulated Data

To demonstrate the efficacy of the proposed methods, we analyzed both simulated and measured data. In the case of simulated data, we tested the adequacy of the introduced concept of a penalty term in the objective function for parameter estimation. The calculated parameter estimates were then compared to the true values. Using the determined parameters, the data were filtered with and without fixed-lag smoothing, and the resulting accuracy was investigated. In addition, the performance of the outlier detection method was quantified.

For the simulated data generation, we defined a model that describes the movement of an object in a 2D plane with a constant turning rate and a constant acceleration. As an example, such a model could be used to filter trajectories in sports as they prevail in ice hockey or on a racing or running track.

The state model included the position in two axes (p_x,p_y), the translational speed v and acceleration a, the turning rate ω and the orientation ϕ (Euler angle). In our example, only the position in 2D was observable. Since, in the case of trajectory filtering in sports, based on position measurements, no control input u_k is available, it will be ignored in the following.

Two datasets $D1$ and $D2$ with different process noise parameters were created. The process noise was introduced via unmodeled independent changes in a and ω (Dataset $D1$) or through v, a, ω and ϕ (Dataset $D2$). The state prediction model can be described in compact form as:

$$\begin{bmatrix} p_{x,k+1} \\ p_{y,k+1} \\ v_{k+1} \\ a_{k+1} \\ \phi_{k+1} \\ \omega_{k+1} \end{bmatrix} = \begin{bmatrix} p_{x,k} + T_k(v_k + a_k T_k) \cdot \cos(\phi_k + \omega_k T_k) \\ p_{y,k} + T_k(v_k + a_k T_k) \cdot \sin(\phi_k + \omega_k T_k) \\ v_k + a_k T_k \\ a_k \\ \phi_k + \omega_k T_k \\ \omega_k \end{bmatrix} \tag{22}$$

$$Q = diag[0\ 0\ \sigma_v^2\ \sigma_a^2\ \sigma_\phi^2\ \sigma_\omega^2] \tag{23}$$

$$R = diag[\sigma_x^2\ \sigma_y^2] \tag{24}$$

The measurement noise was simulated with a standard deviation of σ2x=5·10−3m2 in the x-direction and σ2y=5·10−2m2in the y-direction. At randomly selected times, outliers were added by changing σ2x and σ2y to 50 m^2. The portion of outliers was 5%. To mimic non-uniform sampling, the input time vector was simulated using a Poisson random walk as T_k = 10 ms +1.5 · τ · 10^{-3}, where $\tau \sim$ *Pois*(20).

Each of the simulated datasets consisted of 100 segments of a 30-s length (≈2200 samples each) and was divided into disjoint training (5 segments) and testing subsets (95 segments).

The suitability of the introduction of a penalty term as described in Equation (19) was tested. To achieve the desired effect on the objective function, the penalty term β (Θ) for a given parameter set needs to be bigger than the Mahalanobis distanced_k of valid measurements, thus penalizing the rejection of useful information. In addition, β(Θ) needs to be smaller than the Mahalanobis distance of the outliers to encourage the dismissal of corrupted observations. Therefore, we calculated the percentage of valid measurements for which $\beta(\Theta) > d_k$ and the percentage of outliers for which $\beta(\Theta) < d_k$ as figures of merit (see Table 1). Since the proposed concept also needs to be valid when the correct parameters have not been found yet, we tested all segments of Datasets *D*1 and *D*2 with 1000 randomly chosen combinations of noise parameters. Each single parameter was varied between 10^{-2} times and 10^{2} times its true value, assuming poor *a priori* knowledge of the system.

Table 1. Relation between penalty term β(Θ) and Mahalanobis distances d_k.

	$\beta(\Theta) > d_k,\ k \in M_v$ (%)		$\beta(\Theta) < d_k,\ k \notin M_v$ (%)	
	Mean	**Worst**	**Mean**	**Worst**
*D*1	86.75	78.87	99.21	39.29
*D*2	87.07	77.95	99.59	20.69

The noise parameters and the threshold for outlier detection were estimated. This was done using only the training sets. To evaluate the filtering results, only the test sets were used. The parameter estimation and filtering were conducted using the proposed method, *i.e.*, with consideration of outliers during parameter estimation and filtering. Here, the objective function (20) was used. This approach will be referred to as Method *A*. For comparison, parameter estimation and filtering were also conducted without consideration of outliers. Here, the objective function (18) was used. This approach will be referred to as Method *B*.

In both cases, minimization of the objective function by means of SQP yielded no satisfactory results, since in most cases, it did not converge to a global minimum, even when data without outliers were used, and the initial point was chosen close to the true values. Instead, global optimization with GA was employed.

The estimated parameters were compared to the true values. The results for the simulated datasets can be found in Tables 2 and 3. The optimal Mahalanobis distance thresholds for the detection of outliers were estimated to be $\alpha_1 = 41.6$ for *D*1 and $\alpha_2 = 10.8$ for *D*2.

Table 2. True and estimated noise parameters of Dataset *D*1.

	σ_x^2	σ_y^2	σ_a^2	σ_ω^2
true	5.0×10^{-3}	1.6×10^{-2}	8.2×10^{-2}	5.0×10^{-3}
A	5.6×10^{-3}	1.8×10^{-2}	1.7×10^{-2}	1.4×10^{-3}
B	5.0×10^{-1}	1.6	7.2	3.9×10^{-1}

Table 3. True and estimated noise parameters of Dataset *D*2.

	σ_x^2	σ_y^2	σ_v^2	σ_a^2	σ_ϕ^2	σ_ω^2
true	5.0×10^{-3}	1.6×10^{-2}	1.5×10^{-4}	6.4×10^{-2}	2×10^{-4}	7.3×10^{-3}
A	5.1×10^{-3}	1.6×10^{-2}	9.5×10^{-4}	3.3×10^{-2}	3.7×10^{-3}	5.1×10^{-3}
B	4.9×10^{-1}	1.6	1.4×10^{-2}	5.0	1.4×10^{-2}	6.6×10^{-2}

Using the estimated parameters, the simulated data from the testing sets were filtered with the proposed modified EKF. Moreover, the root-mean-square-error (RMSE) between the simulated ground truth data and the filtered output was calculated for time lags between 0 s and 0.5 s. As a reference, the simulated datasets were also created without outliers and filtered using the true noise parameters. The results are depicted in Figures 4 and 5.

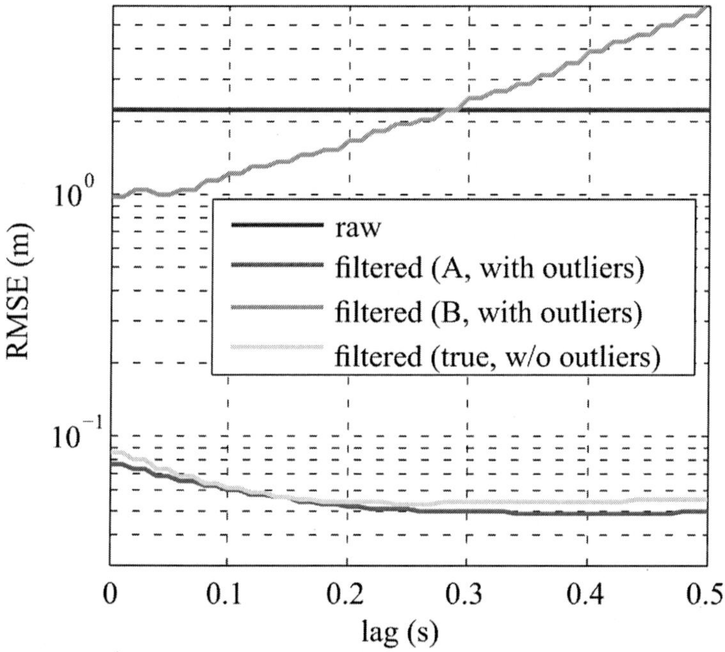

Figure 4. RMSE of Dataset *D*1 (without training set). The blue line shows the RMSE of the unfiltered data (including outliers). The RMSE obtained by filtering the data (including outliers) with the parameters from the proposed method (*A*) is depicted in purple. The orange curve corresponds to the filtering results obtained using the parameter estimation from [34], without special handling of the outliers in the data (*B*). The yellow line represents the results for filtering the data without outliers and with the true simulated noise parameters. The ripple in the filtered curves corresponds to the smallest sampling interval of 10 ms.

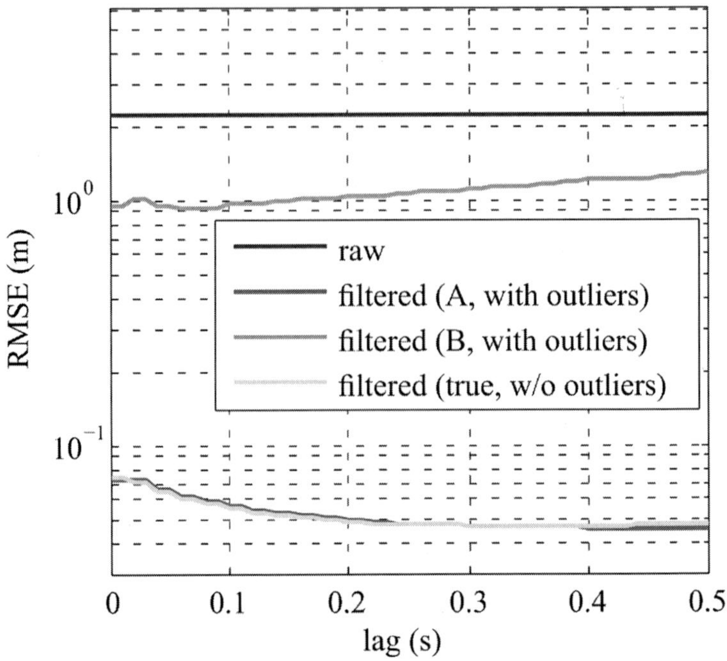

Figure 5. RMSE of Dataset *D*2 (without training set). The results were obtained in analogy to those from Dataset *D*1, depicted in Figure 4.

Table 4. Accuracy of the proposed outlier detection method.

	Sensitivity (%)		Specificity (%)	
	Mean	Worst	Mean	Worst
*D*1	99.86	96.67	99.53	96.62
*D*2	99.27	94.12	99.99	99.86

To evaluate the accuracy of the proposed outlier detection method, sensitivity and specificity were calculated for all segments of Datasets *D*1 and *D*2. The results are summarized in Table 4.

3.2. Real Data

In addition to the tests with simulated data, the filtering approach was also applied to real data. It consisted of 2D position measurements from a local positioning system (LPS). The LPS employed the frequency-modulated continuous-wave (FMCW) principle. It was comprised of 17 base stations and a reference transponder with fixed and known positions. FMCW signals sent from the mobile target transponder and the reference transponder were used to obtain several time differences of arrival (TDOA) at each sampling time. From these TDOA measurements, the positions of the target were acquired with a non-uniform sampling rate. The time interval between subsequent samples ranged from 5 ms to 400 ms, and the measurements were corrupted by outliers. The LPS was used to observe the movements of a quad bike on a race track. The analyzed LPS data contained approximately 40,000 samples.

To get ground truth measurements for comparison and evaluation, the position of the quad was simultaneously recorded using a Real-Time Kinematic Global Positioning System (RTK-GPS). The measurements from this carrier-phase enhanced differential GPS were evaluated using RTKLIB [44]. The reference antenna for the RTK-GPS was positioned within a distance of less than 1 km from the mobile receiver at all times. Thus, the RMSE of the reference measurements can be assumed to be smaller than 1 cm [45]. The RTK-GPS positions were measured with a constant sampling rate of 5 Hz.

The LPS data were filtered using the presented methods. The output of the filter was synchronized to the reference measurements to allow comparison.

The estimation of noise parameters and the filtering has been performed both with Method *A* and *B*, with different delays, as described in Section 3.1. A plot of an extract of the measured, filtered (500-ms lag) and ground truth data is depicted in Figure 6.

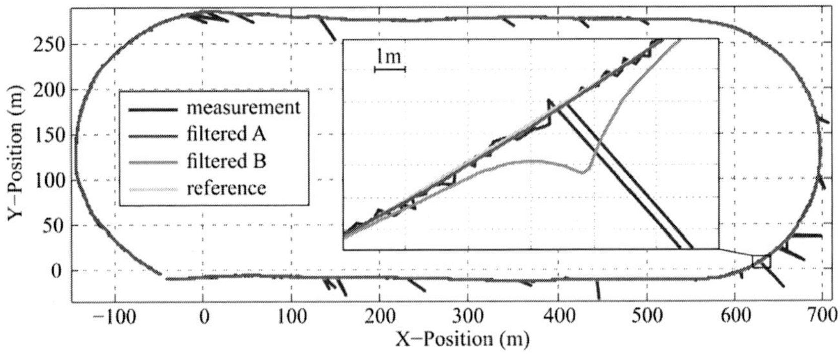

Figure 6. Example plot of the measured and filtered data with magnified detail of the lower right corner of the track. The filtered curves were processed using a constant lag of 500 ms. The measurements were disturbed by Gaussian noise and outliers. Using the proposed method (*A*), the Gaussian noise was reduced, and there was no visible effect of the outliers in the observations. When outliers were not considered in the filtering process (*B*), the filter is susceptible to the outliers, and the filtered curve deviated from the reference. In the depicted case, the deviation added up to several meters.

Since the true noise parameters are not known in this case, they cannot be resorted to for evaluation. Instead, the RMSE serves as the figure of merit. The results in terms of RMSE are visualized in Figure 7.

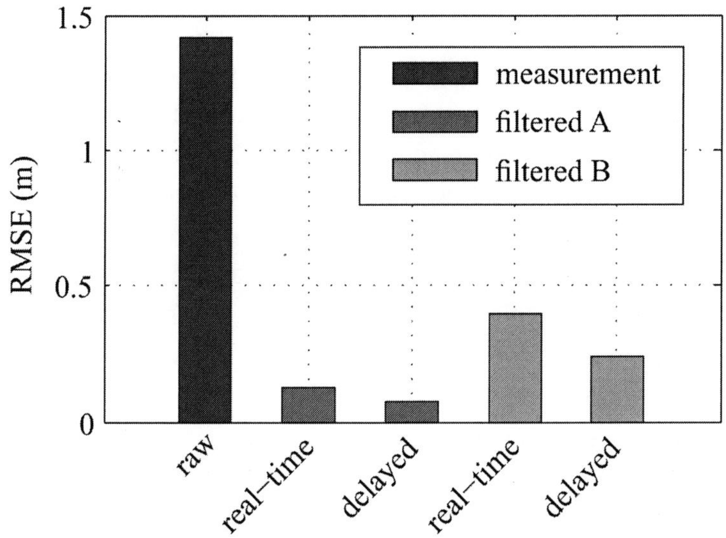

Figure 7. RMSE for filtering real data. The RMSE with respect to the RTK-GPS reference data adds up to 1.42 m for the raw measurements, 0.13 m for the filtered positions (no lag) according to Method *A* and 0.40 m for Method *B*. Using delayed filtering with a constant lag of 500 ms, the RMSE was reduced to 0.07 m for Method A and to 0.24 m for Method *B*.

4. DISCUSSION

4.1. Simulated Data

The experimental results based on the simulated datasets showed that the proposed concept for the consideration of outliers in the parameter estimation works with sufficient reliability. The introduction of a penalty term mitigates the effect of outliers, even when only poor estimates of the parameters were assumed. The rejection of outliers led to an improvement of the objective function in almost all cases, which compensated for their negative effect. Although there were some cases where the objective function was corrected for only a small fraction of the outliers (worst case: 20.69%), valid parameter estimates could still be found, due to the following two reasons: First, in only 0.24% of the analyzed 200,000 runs, the fraction of outliers that were corrected in the objective function was below 90%. Second, in these cases, the resulting value of the objective function was approximately five-times higher than for the rest, *i.e.*, these parameter sets were far away from the optimal parameters and would be discarded by the optimization algorithm. In the case of valid observations, the objective function was not influenced optimally in around 13% of the runs. This could slow down the convergence of the optimization algorithm.

Using the proposed optimization as expressed in Equation (20), appropriate estimates of the noise parameters in the presence of outliers were determined. The estimated measurement noise parameters acquired with the proposed method (*A*) matched the true values almost exactly, whereas the process noise parameters showed some deviations from the ground truth in both investigated datasets. The noise parameters estimated without special treatment of the outliers (*B*) were considerably overestimated in all cases and differed from the true values by up to two magnitudes. Thus, the negligence of the outliers in the parameter estimation led to an unacceptable overestimation of the real noise parameters.

A drawback of the proposed estimation of the noise parameters by means of a global optimization algorithm, like GA, is the fact that it can become computationally expensive in the case of high order systems. Since convergence to the global optimum is not guaranteed, many function evaluations are usually necessary to find a suitable solution. However, parametrization of the noise covariance matrices as diagonal instead of full matrices can reduce the computational effort in the case of independent noise sources.

Regarding the accuracy of the filtered output, Figures 4 and 5 show the efficacy of the proposed methods: even when non-uniformly-sampled data containing outliers were to be filtered with a constant lag (purple curves), the results in terms of RMSE are as good as if there were no outliers at all in the data (yellow curves). In the case of Dataset *D*1 the RMSE using the proposed filter was actually slightly better than for the reference (which contains no outliers). This is probably caused by the fact that not only the artificially-introduced outliers were rejected, but also some samples that were overly corrupted by the normal measurement noise. This is equivalent to cutting off the tails of the underlying Gaussian distribution and thus, limits the measurement error leading to better filtering results.

The accuracy of the filtered results increased with bigger lags when using the proposed method. This also required increased computation time, since with bigger lags more observations were included in the calculation of each output value.

Finally, the proposed outlier detection method performed robustly. Considering a sensitivity and specificity of more than 94% in all analyzed cases, we conclude that outliers can be found with high accuracy.

4.2. Real Data

The results based on real data support the outcomes acquired from the filtering of simulated data. Using the proposed methods, the measured positions were filtered robustly, and clear improvements in terms of RMSE could be achieved.Figure 6 shows that there is no visible effect of outliers on the filtered trajectory for the proposed method (*A*), whereas the filtered curve deviates considerably from the reference measurements when outliers are not considered in parameter estimation and filtering (*B*). Although it is argued in [46] that the hard rejection of outliers is suboptimal, this approach has shown robust operation in the presented simulated and real scenarios, which is in line with other findings in the literature [12,14,15]. The proposed filtering procedure is also compatible with other approaches for the treatment of outliers. Therefore, the results might be further improved, if other robustification methods, (e.g., as described in [17]) are employed in conjunction with the proposed framework.

The use of the presented method for delayed filtering led to further improvements of the filtered signal in terms of RMSE. The fixed-lag filtering allowed enhanced filtering results with a constant time delay and output rate, although the LPS position data were sampled non-uniformly. In contrast with the findings from the simulated datasets, the RMSE decreased even for Method *B* when fixed-lag smoothing was employed. This is presumably due to the smaller amount of outliers in the real measurement data compared to the simulated data; thus, the effect of the outliers was smaller. However, the resulting errors of the filtered output using the presented methods (*A*) was significantly smaller, both with and without smoothing. In this case, the RMSE of the measurements was reduced by 91% without smoothing and 95% with smoothing.

These findings demonstrate that the proposed filtering approach is also applicable under realistic conditions and that considerable enhancements can be achieved.

5. CONCLUSIONS

In this paper, we introduced a novel Kalman-based filtering framework. It is comprised of several features, namely the handling of non-linear systems, non-uniform sampling and outliers, while at the same time allowing optimal utilization of available information by fixed-lag smoothing and automated parameter estimation. Up to now, only small subsets of these features could be employed at the same time, and the advantages of the others had to be relinquished due to their irreconcilability.

The results based on simulated data exemplify that even in the presence of outliers and with non-uniformly-sampled measurements, suitable noise parameters of linear or non-linear systems can be estimated, and the filtered output can be calculated in real time or with a constant delay. The automated noise estimation makes cumbersome manual parameter tuning redundant. Further, the error of the filtered output is less or equal to the error that would result from filtering data without outliers and with perfectly known system characteristics.

Experimental results based on real measurements testify to the practicability of the newly developed filter in realistic situations and its potential for substantial enhancements of the measured data.

Future work includes investigations on whether and how the parameter estimation can be sped up to increase the viability in case of high-dimensional systems. A robustification of the EM-algorithm for parameter estimation might be a possible solution to reduce the computational cost. Moreover, the transfer of the presented combination of modifications to the UKF and SRUKF will be the subject of future investigations.

With the methods presented in this paper, multiple extensions of the KF become applicable in situations where they could not be used so far, given that the pre-conditions were not met. Thus, the newly presented method expands the boundaries for the application of Kalman-based filtering, which make its advantages accessible even under challenging conditions.

ACKNOWLEDGMENTS

The authors would like to thank Cristian Pasluosta for his valuable suggestions and support.

AUTHOR CONTRIBUTIONS

Thomas Kautz designed the experiment and performed the statistical analysis. Thomas Kautz and Bjoern M. Eskofier conceived of the study, and drafted the manuscript. Both authors approved the manuscript.

CONFLICTS OF INTEREST

The authors declare no conflict of interest.

REFERENCES

1. Kalman, R. A new approach to linear filtering and prediction problems. *J. Basic Eng.* **1960**, *82*, 35–45.

2. Boley, D.; Sutherland, K. Recursive total least squares: An alternative to the discrete Kalman filter., Proceedings of the AMS Meeting Applied Linear Algebra, DeKalb, IL, USA, May 1993; pp. 1–17.

3. Ljung, L. Asymptotic behavior of the extended Kalman filter as a parameter estimator for linear systems. *IEEE Trans. Autom. Control* **1979**, *24*, 36–50.

4. Moore, J. Discrete-time fixed-lag smoothing algorithms. *Automatica* **1973**, *9*, 163–173.

5. Liu, H.; Shah, S.; Jiang, W. On-line outlier detection and data cleaning. *Comput. Chem. Eng.* **2004**, *28*, 1635–1647.

6. Welch, G.; Bishop, G.; Hill, C. *An Introduction to the Kalman Filter*; University of North Carolina: Chapel Hill, NC, USA, 1995; pp. 1–16.

7. Julier, S.J.; Uhlmann, J.K. A New Extension of the Kalman Filter to Nonlinear Systems. *Int. Symp. Aerosp./Def. Sens. Simul. Controls* **1997**, *3*, 182–193.

8. Van der Merwe, R.; Wan, E.A. The Square-Root Unscented Kalman Filter for State and Parameter-Estimation. Proceedings of the 2001 IEEE International Conference on Acoustics, Speech, and Signal Processing, 2001. Proceedings. (ICASSP '01), Salt Lake City, UT, USA, 7–11 May 2001; Volume 6.

9. Hürzeler, M.; Künsch, H.R. Monte Carlo approximations for general state-space models. *J. Comput. Graph. Stat.* **2001**,*7*, 175–193.

10. Grubbs, F.E. Procedures for detecting outlying observations in samples. *Technometrics* **1969**, *11*, 1–21.

11. Agamennoni, G.; Nieto, J.I.; Nebot, E.M. An Outlier-Robust Kalman Filter. Proceedings of the 2011 IEEE International Conference on Robotics and Automation (ICRA), Shanghai, China, 9–13 May 2011; pp. 1551–1558.

12. Ting, J.-A.; Theodorou, E.; Schaal, S. A Kalman filter for robust outlier detection. Proceedings of the IROS 2007IEEE/RSJ International Conference on Intelligent Robots and Systems, San Diego, CA, USA, 29 October– 2 November 2007; pp. 1514–1519.

13. Gandhi, M.A.; Mili, L. Robust Kalman filter based on a generalized maximum-likelihood-type estimator. *IEEE Trans. Signal Process.* **2010**, *58*, 2509–2520.

14. Meyr, H.; Spies, G. The structure and performance of estimators for real-time estimation of randomly varying time delay. *IEEE Trans. Acoust. Speech Signal Process.* **1984**, *32*, 81–94.

15. Mirza, J.M. A Modified Kalman Filter for Non-gaussian Measurement Noise. In *Communication Systems and Information Technology*; Springer: Berlin/Heidelberg, Germany, 2011; pp. 401–409.

16. Schick, I.; Mitter, S. Robust recursive estimation in the presence of heavy-tailed observation noise. *Ann. Stat.* **1994**, *22*, 1045–1080.

17. Ruckdeschel, P. Ansätze zur Robustifizierung des Kalman-Filters. *Bayreuther Math. Schriften* **2001**, *64*, 21–33.

18. Ruckdeschel, P.; Spangl, B.; Pupashenko, D. Robust Kalman tracking and smoothing with pro-pagating and non-propagating outliers. *Stat. Pap.* **2014**, *55*, 93–123.

19. Stockinger, N.; Dutter, R. Robust time series analysis: A survey. *Kybernetica* **1987**, *23*, 3–88.

20. Ershhov, A.A.; Lipster, R.S. Robust Kalman Filter in Discrete Time. *Autom. Remote Control* **1987**, *39*, 359–367.

21. Martin, R.D.; Raftery, A.E. Comment: Robustness, Computation, and Non-Euclidean Models. *J. Am. Stat. Assoc.* **1987**,*82*, 1044–1050.

22. Armesto, L.; Tornero, J. SLAM based on Kalman filter for multi-rate fusion of laser and encoder measurements. Proceedings of the 2004 IEEE/RSJ International Conference on Intelligent Robots and Systems 2004 (IROS 2004), Proceedings. Chicago, IL, USA, 28 September–2 October 2004; Volume 2, pp. 1860–1865.

23. Sridhar, B.; Smith, P. Multirate and event-driven Kalman filters for helicopter flight. *IEEE Control Syst.* **1993**, *13*, 26–33.

24. Li, W.; Shah, S.L.; Xiao, D. Kalman filters in non-uniformly sampled multirate systems: For FDI and beyond.*Automatica* **2008**, *44*, 199–208.

25. Simon, D. *Optimal State Estimation: Kalman, H∞, and Nonlinear Approaches*; John Wiley & Sons, Inc.: Hoboken, NJ, USA, 2006; pp. 263–296.

26. Vauhkonen, P.J.; Vauhkonen, M.; Kaipio, J.P. Fixed-lag smoothing and state estimation in dynamic electrical impedance tomography. *Int. J. Numer. Meth. Eng.* **2001**, *50*, 2195–2209.

27. Cohn, S.; Sivakumaran, N.; Todling, R. A fixed-lag Kalman smoother for retrospective data assimilation. *Mon. Weather Rev.* **1994**, *122*, 2838–2867.

28. Chen, B.; Tugnait, J. Tracking of multiple maneuvering targets in clutter using IMM/JPDA filtering and fixed-lag smoothing. *Automatica* **2001**, *37*, 239–249.

29. Wang, J.; Gopaul, S.; Guo, J. Adaptive Kalman Filtering Based on Posteriori Variance-Covariance Components Estimation. Proceeding of the CPGPS Technical Forum Satellite Navigation and Positioning, Shanghai, China, 17–20 August 2010; pp. 1–11.

30. Ghahramani, Z.; Hinton, G. *Parameter Estimation for Linear Dynamical Systems*; Technical Report; Department of Computer Science, University of Totronto: Toronto, ON, Canada, 1996.

31. Axelsson, P.; Orguner, U. ML estimation of process noise variance in dynamic systems. Proceedings of the 18th IFAC World Congress, Milano, Italy, 28 August–2 September 2011; pp. 5609–5614.

32. Dempster, A.P.; Laird, N.M.; Rubin, D. Maximum likelihood from incomplete data via the EM algorithm. *J. R. Stat. Soc.* **1977**, *3*, 1–38.

33. Shumway, R.H.; Stoffer, D.S. An approach to time series smoothing and forecasting using the EM algorithm. *J. Time Ser. Anal.* **1982**, *3*, 253–264.

34. Bavdekar, V.A.; Deshpande, A.P.; Patwardhan, S.C. Identification of process and measurement noise covariance for state and parameter estimation using extended Kalman filter. *J. Process Control* **2011**, *21*, 585–601.

35. Kassam, S.A.; Poor, H.V. Robust techniques for signal processing: A survey. *IEEE Proc.* **1985**, *73*, 433–4814.

36. Fox, A.J. Outliers in time series. *J. R. Stat. Soc. Ser. B* **1972**, *34*, 350–363.

37. Tukey, J.W. *Exploratory Data Analysis*; Addison-Wesley: Reading, MA, USA, 1977.

38. Julier, S.; Uhlmann, J.; Durrant-Whyte, H. A new approach for filtering nonlinear systems. Proceedings of the 1995 American Control Conference, - ACC'95, Seattle, WA, USA, 21–23 June 1995; Volume 3, pp. 1628–1632.

39. De Geeter, J.; van Brussel, H.; de Schutter, J.; Decéton, M. A smoothly constrained Kalman filter. *IEEE Trans. Pattern Anal. Mach. Intell.* **1997**, *19*, 1171–1177.

40. Rauch, H.E.; Striebel, C.T.; Tung, F. Maximum likelihood estimates of linear dynamic systems. *AIAA J.* **1965**, *3*, 1445–1450.

41. Lawrence, D. *Handbook of Genetic Algorithms*; Lawrence, D., Ed.; Van Nostrand Reinhold: New York, NY, USA, 1991.

42. Willis, J.; Hiden, H.G.; Marenbach, P.; McKay, B.; Montague, G.A. Genetic programming: An introduction and survey of applications. Proceeding of Genetic Algorithms in Engineering Systems: Innovations and Applications, Glasgow, UK, 2–4 September 1997; pp. 314–319.

43. Herrera, E.P.; Kaufmann, H. Adaptive Methods of Kalman Filtering for Personal Positioning Systems. Proceedings of the 23rd International Technical Meeting of the Satellite Division of the Institute of Navigation, Portland, OR, USA, 21–24 September 2010; pp. 584–589.

44. Takasu, T. *RTKLIB: Open Source Program Package for RTK-GPS*; FOSS4G: Jeju, Korea, 2009.

45. Takasu, T.; Kubo, N.; Yasuda, A. *Development, Evaluation and Application of RTKLIB: A Program Library for RTK-GPS*; Int. GPS/GNSS Symposium: Jeju, Korea, 2007.

46. Ruckdeschel, P. *Optimally robust Kalman Fltering*; Technical Report; Fraunhofer ITWM Kaiserslautern: Kaiserslautern, Germany, 2010.

CHAPTER 9

Modified Particle Filtering Algorithm for Single Acoustic Vector Sensor DOA Tracking

Xinbo Li [1], Haixin Sun [2,†], Liangxu Jiang [1,†], Yaowu Shi [1,†] and Yue Wu [3,*]

[1] School of Communication Engineering, Jilin University, Renmin Street No. 5988, Changchun 130022, China

[2] School of Electronic and Information Engineering, Changchun University, Weixing Road, No. 6543, Changchun 130022, China

[3] School of Mechanical Science and Engineering, Jilin University, Renmin Street No. 5988, Changchun 130022, China

ABSTRACT

The conventional direction of arrival (DOA) estimation algorithm with static sources assumption usually estimates the source angles of two adjacent moments independently and the correlation of the moments is not considered. In this article, we focus on the DOA estimation of moving sources and a modified particle filtering (MPF) algorithm is proposed with state space model of single acoustic vector sensor. Although the particle filtering (PF) algorithm has been introduced for acoustic vector sensor applications, it is not suitable for the case that one dimension angle of source is estimated with large deviation, the two dimension angles (pitch angle and azimuth angle) cannot be simultaneously employed to update the state through resampling processing of PF algorithm. To solve the problems mentioned above, the MPF algorithm is proposed in which the state estimation of previous moment is introduced to the particle sampling of present moment to improve the importance function. Moreover, the independent relationship of pitch angle and azimuth angle is considered and the two dimension angles are sampled and evaluated, respectively. Then, the MUSIC spectrum function is used as the "likehood" function of the MPF algorithm, and the modified PF-MUSIC (MPF-MUSIC) algorithm is proposed to improve the root mean square error (RMSE) and the probability of convergence. The theoretical analysis and the simulation results validate the effectiveness and feasibility of the two proposed algorithms.

KEYWORDS

DOA tracking; particle filtering; importance function; acoustic vector sensor

1. INTRODUCTION

Acoustic vector sensor is a new kind of sensors that can obtain the information of the scalar field and the vector field of acoustic wave simultaneously. Its application covers the areas of mobile positioning, sonar system, radar tracking, fault source location, *etc*. Since single acoustic vector sensor is a simplified version of the acoustic vector array and its array aperture can be expanded for multiple sources location estimation, some related issues have been considered

[1,2,3,4,5,6,7]. The near-field manifold of single acoustic vector sensor has been derived in [1]. Then, the azimuth-elevation direction of arrival plus radial distance estimation algorithm for near-field wideband emitter was developed in [2]. The beamforming and blind interference rejection algorithm for fast frequency-hop spread spectrum signals was presented in [3]. The Cramer-Rao bound for single acoustic vector sensor DOA estimation with model errors was investigated in [4,5]. The spatial collocation requirement of single acoustic vector sensor was relaxed and four-component sensors were located separately, which realizes an extended spatial aperture [6,7]. Usually, the DOA estimation algorithm based on acoustic vector sensor adopts multiple snapshot and batch processing to estimate the target parameters, which is also named as static DOA estimation [1,2,3,4,5,6,7,8,9,10]. However, in the real environment, the target source is moving and the batch processing of the static DOA estimation algorithm cannot achieve the continuous tracking. Furthermore, the static estimation algorithm estimates the angles of two different moments independently and does not consider the correlation of the moments, which leads to the large dynamic estimation deviation. The conventional static DOA estimation algorithm cannot fulfill the increasing accuracy and real-time demands. Hence, the study of DOA tracking estimation for dynamic sources becomes a new trend.

To handle the DOA tracking for dynamic sources, several non-stationary adaptive algorithms have been proposed, such as matched-field processing [11], H_∞ tracking [12], Kalman filtering [13], adaptive recursive-least-squares tracking [14,15] particle filtering (PF) [16,17,18,19,20], *etc.* The objectives of these algorithms are to improve the dynamic localization performance. Since particle filtering can deal with any dynamic system of linear Gaussian or nonlinear non-Gaussian represented by state space models [21,22,23,24,25,26] and the observation equation of the acoustic vector sensor is a highly nonlinear function of state, the PF algorithm has been introduced to the field of the acoustic vector array signal processing. In [17], the single acoustic vector sensor DOA tracking algorithm was proposed with particle filtering, the maximum "likehood" function was used as the observation likehood function and the Cramer-Rao bound for dynamic DOA estimation under the Gaussian background was derived in details. In [18], the CAPON and Barlett beamforming spectrum functions were used as the likehood function of the PF algorithm and the modified particle filtering DOA tracking algorithm with the exponential factor weighting was proposed. In [19], the algorithms [17,18] were adopted and verified in the real marine environment and the effectiveness and anti-jamming capability of different algorithms were validated. The final results indicated that the algorithm proposed in [18] outperformed other algorithms in the real marine noise environment. The algorithms [17,18] were established under the conditions of Gaussian background noise. In [20], the DOA tracking under the impulsive noise was studied. The observation likehood function based on fractional order covariance matrix was derived and the combination of the fractional low order of moment and the PF (FLOM-PF) algorithm was proposed.

The study results mentioned above presented that the CAPON algorithm is affected by the number of array elements. When the element number is small, the estimation deviation is large. However, the MUSIC algorithm had better resolution performance and sharper spectrum peak when the distance of the targets is close [27]. The MUSIC algorithms for acoustic vector sensor applications have been presented in [28,29,30]. As the likehood function of the particle filtering, the spectrum function of the MUSIC algorithm has better evaluation for the particles instead of the spectrum functions of CAPON algorithm or Barlett beamforming algorithm.

On the basis of previous studies, the contents of this paper are as follows. Firstly, considering the independent relationship of pitch angle and azimuth angle, the two angles are sampled and evaluated independently to avoid large deviation of either pitch angle or azimuth angle, which leads to the other angle being unable to update the state accurately, through resampling, and a modified PF algorithm is proposed which improves the importance probability density.

Secondly, the spectrum function of MUSIC algorithm is exploited as the likehood function of PF, instead of the spectrum functions of CAPON algorithm and Barlett beamforming algorithm, and the modified PF-MUSIC tracking algorithm is proposed. Finally, the feasibility and effectiveness of the algorithms are validated through simulation results.

2. STATE SPACE MODEL

2.1. The Establishment of the State Equation

Since considering multiple acoustic sources existing simultaneously is more complex than a single acoustic source, we mainly consider the single acoustic source issue. Assuming the moving speed of DOA is constant, the variation of the source angle is simulated by the two order constant velocity model and the state equation of single angle is established. The case of multiple state variables can be derived in a similar way.

Given that the source angle is represented by θ moving along a straight line in a certain period of time and the sampling time is ΔT, at the sampling point $k\Delta$T, the real pitch angle of the target $\theta(k)$, the moving speed is θ·(k). According to the equation of the constant velocity motion, we have

$$\theta(k+1)=\theta(k)+\dot{\theta}(k)\Delta T+\frac{1}{2}\Delta T^2u(k) \tag{1}$$

$$\dot{\theta}(k+1)=\dot{\theta}(k)+\Delta Tu(k) \tag{2}$$

where $u(k)$ is the speed fluctuations caused by external factors such as friction force or wind [19]. Assuming that the noise is zero-mean

Gaussian random noise and independent with the observation noise, the state space model of variable θ can be derived as,

$$\begin{bmatrix} \theta(k+1) \\ \dot{\theta}(k+1) \end{bmatrix} = \begin{bmatrix} 1 & T_0 \\ 0 & 1 \end{bmatrix} \begin{bmatrix} \theta(k) \\ \dot{\theta}(k) \end{bmatrix} + \begin{bmatrix} \frac{1}{2}\Delta T^2 \\ \Delta T \end{bmatrix} u(k) \tag{3}$$

Similarly, consider the case of the source state expanded to two angles. In the k moment, the four-dimensional state vector of the source is represented by xk=[θk,θk·,φk,φk·]^T, where θ_k is the pitch angle, θ·k is the velocity variable of pitch angle,φ_k is the azimuth angle and φ·k is the velocity variable of azimuth angle. Hence, the state equation of the acoustic source angle is

$$x_k = Ax_{k-1} + Bu_k \tag{4}$$

where u_k is zero-mean Gaussian random noise, A is the state coefficient matrix and B is the input coefficient matrix. For the DOA estimation and tracking of the acoustic vector sensor, the value of A and B are

$$A = \begin{bmatrix} 1, \Delta T, 0, 0 \\ 0,\ 1,\ 0, 0 \\ 0, 0, 1, \Delta T \\ 0, 0, 0,\ 1 \end{bmatrix}, \quad B = \begin{bmatrix} \Delta T^2/2 \\ \Delta T \\ \Delta T^2/2 \\ \Delta T \end{bmatrix} \tag{5}$$

2.2. The Establishment of the Observation Model

Assuming a far-field narrowband signal with λ_s wavelength impinges into a single acoustic vector sensor with impinging angles Θs=(θs,φs), where φs∈(0,2π) is the pitch angle and θs∈(−π/2,π/2) is the azimuth angle of the impinging signal, respectively. The sensor location in the coordinates is r=(x0,y0,z0). The sound pressure of the source is shown as*s*(*t*), which is an unknown deterministic signal. The unit vector u⃗ s is (sinθscosφs,sinθssinφs, cosθs), which is pointed from the origin to the direction of the acoustic source. The vibration velocities of the acoustic source on the x–y–z coordinate axes are represented by vx(t),vy(t),vz(t), respectively. According to the relationship of the particle vibration velocities and the sound pressure, we have:

$$\begin{aligned} v_x(t) &= \sin\theta_s \cos\varphi_s * s(t) \\ v_y(t) &= \sin\varphi_s \sin\theta_s * s(t) \\ v_z(t) &= \cos\theta_s * s(t) \end{aligned} \tag{6}$$

The wave delay of the acoustic source in the present sensor element is

$$\Phi_s = \frac{2\pi}{\lambda}(x_0 \cos\theta_s \sin\varphi_s + y_0 \sin\theta_s \sin\varphi_s + z_0 \cos\varphi_s) \tag{7}$$

The output of the array element is represented by

$$y(t) = a(\Phi_s) \otimes [1, \cos\theta_s \sin\varphi_s, \sin\theta_s \sin\varphi_s, \cos\varphi_s]^T s(t) + n(t) \tag{8}$$

where y(t)⊂C4*1, [·]T represents the transpose operation, ⊗ represents the direct product, *n*(*t*) is the interference noise at the receiving end, and a(Φs) is the sensor response coefficient to the acoustic source, and it is:

$$a(\Phi_s) = \exp(-j\Phi_s) \tag{9}$$

In the practical engineering application, to guarantee the real-time and effectiveness of the estimation, the observation matrix of the array is obtained with the limited number of snapshots. Assuming *L* observations at the *k* moment, the receiving matrix is:

$$\mathbf{Y}(k) = a(\Phi_s)[1, \cos\theta_s \sin\varphi_s, \sin\theta_s \sin\varphi_s, \cos\varphi_s]^T \otimes s(k) + \mathrm{N}(k) \\ = b(\Phi_s) \otimes s(k) + \mathrm{N}(k) \tag{10}$$

where

$$b(\Phi_s) = a(\Phi_s)[1, \cos\theta_s \sin\varphi_s, \sin\theta_s \sin\varphi_s, \cos\varphi_s]^T \tag{11}$$

$$s(k) = \left[s(kL+1), s(kL+2), \cdots, s(kL+L) \right] \tag{12}$$

3. MODIFIED PARTICLE FILTERING TRACKING ALGORITHM

3.1. Particle Filtering Algorithm

The Bayes importance sampling algorithm [19] can be achieved with sequential importance sampling (SIR) process in recursive way and the state filtering estimation can be accomplished in predicting and updating way. The prediction process is

$$p\left(x_k \middle| \mathbf{Y}_{1:k-1}\right) = \int p\left(x_k \middle| x_{k-1}\right) p\left(x_{k-1} \middle| \mathbf{Y}_{1:k-1}\right) dx_{k-1} \tag{13}$$

where p(·) represents the conditional probability. The update process is

$$p\left(x_k \middle| \mathbf{Y}_{1:k}\right) = \frac{p\left(\mathbf{Y}_k \middle| x_k\right) p\left(x_k \middle| \mathbf{Y}_{1:k-1}\right)}{\int p\left(\mathbf{Y}_k \middle| x_k\right) p\left(x_k \middle| \mathbf{Y}_{1:k-1}\right) dx_k} \tag{14}$$

Considering the relationship of the two adjacent observing time, the sequential importance sampling process calculates the sampling weights through recursion. At the k moment, the updated equation of the weight of the particle $x_k(i)$ is

$$w_k^*\left(x_k(i)\right)=w_{k-1}^*\left(x_{k-1}(i)\right)\frac{p\left(\mathbf{Y}_k\left|x_k(i)\right.\right)p\left(x_k(i)\left|x_{k-1}(i)\right.\right)}{q\left(x_k(i)\left|x_{0:k-1}(i),\mathbf{Y}_{1:k}\right.\right)} \tag{15}$$

where q(xk|x0:k−1,Y1:k) is the importance probability density, xk(i),i=1,⋯,N is the particle samples at the k moment. In the SIR algorithm [23], the simplified p(xk|xk−1) is employed as the importance probability density and the particle samples are extracted randomly from this distribution, which is represented by

$$q\left(x_k(i)\left|x_{0:k-1}(i),\mathbf{Y}_{1:k}\right.\right)=p\left(x_k(i)\left|x_{k-1}(i)\right.\right) \tag{16}$$

The corresponding weight calculating formula of the particles is

$$w_k^*\left(x_k(i)\right)=w_{k-1}^*\left(x_{k-1}(i)\right)p\left(\mathbf{Y}_k\left|x_k(i)\right.\right) \tag{17}$$

The normalized weight coefficient is

$$w_k\left(x_k(i)\right)=\frac{w_k^*\left(x_k(i)\right)}{\sum_{i=1}^{N}w_k^*\left(x_k(i)\right)} \tag{18}$$

Hence, the estimated value of x_k is

$$x_k=\sum_{i=1}^{N}w_k(i)x_k(i) \tag{19}$$

3.2. Modified Particle Filtering Algorithm

The system state includes two parameters, the pitch angle and the azimuth angle of the impinging signal. From the derivation of the particle filtering algorithm [19], it is known that ($\theta_k(i)$, $\varphi_k(i)$) is used as a state combination for estimation. Hence, the algorithms in [19] may encounter some problems, especially when the pitch angle is in the vicinity of the real state while the azimuth angle has large deviation from the real state, or *vice versa*, the weight $w_k\ (i)$ for the combination of angles ($\theta_k\ (i)$, $\varphi_k\ (i)$) is small relatively. The small weight samples cannot be duplicated and propagated by resampling process. Thus, the effective pitch angle (or azimuth angle) samples in the vicinity of the real state are overlooked due to not passing the resampling process. Hence, in the case of two estimated angles, the two angles should be evaluated independently. The problems that the expected angles cannot be duplicated and propagated through resampling process with large deviation of pitch angle or azimuth angle can be avoided.

Here, the improvements of the evaluation way for the pitch angle and the azimuth angle are introduced to the SIR particle filtering. The corresponding importance probability density functions of the pitch angle and the azimuth angle are given in Equations (20) and (21), respectively,

$$q\left(\theta_k(i)\middle|\theta_{0:k-1}(i),y_{1:k}\right)=p\left(\theta_k(i)\middle|\theta_{k-1}(i),\hat{\varphi}_{k-1}\right) \tag{20}$$

$$q\left(\varphi_k(i)\middle|\varphi_{0:k-1}(i),y_{1:k}\right)=p\left(\varphi_k(i)\middle|\varphi_{k-1}(i),\hat{\theta}_{k-1}\right) \tag{21}$$

where $\hat{\varphi}_{k-1}$ and $\hat{\theta}_{k-1}$ are the estimated value of the state (pitch angle and the azimuth angle) in the previous moment, respectively. The particles $\left(\theta_k(i),\hat{\varphi}_{k-1}(i)\right)$ and $\left(\varphi_k(i),\hat{\theta}_{k-1}(i)\right)$ are obtained by sampling through the probability density represented in Equations (20) and (21).

The essence of the modified PF algorithm is to construct new particle samples by exploiting the estimated angle in the previous moment, which is equivalent to the expansion of the particle number. Since the particle samples of the pitch angle and the azimuth angle at the k moment are relevant to the samples at the $k-1$ moment and irrelevant to the estimated value at the $k-1$ moment, Equations (20) and (21) satisfy

$$p\left(\theta_k(i)\middle|\theta_{k-1}(i),\hat{\varphi}_{k-1}\right)=p\left(\theta_k(i)\middle|\theta_{k-1}(i)\right) \tag{22}$$

$$p\left(\varphi_k(i)\middle|\varphi_{k-1}(i),\hat{\theta}_{k-1}\right)=p\left(\varphi_k(i)\middle|\varphi_{k-1}(i)\right) \tag{23}$$

Substitute Equations (22) and (23) into Equation (17)

$$w_{k\theta}^{*}(i)=w_{(k-1)\theta}^{*}(i)\,p\left(\mathbf{Y}_k\middle|\theta_k(i),\hat{\varphi}_{k-1}(i)\right) \tag{24}$$

$$w_{k\varphi}^{*}(i)=w_{(k-1)\varphi}^{*}(i)\,p\left(\mathbf{Y}_k\middle|\varphi_k(i),\hat{\theta}_{k-1}(i)\right) \tag{25}$$

After normalization, the weights are represented by:

$$w_{k\theta}(i)=\frac{w_{k\theta}^{*}(i)}{\sum_{i=1}^{N}w_{k\theta}^{*}(i)} \tag{26}$$

$$w_{k\varphi}(i)=\frac{w_{k\varphi}^{*}(i)}{\sum_{i=1}^{N}w_{k\varphi}^{*}(i)} \tag{27}$$

Hence, the estimated values of the impinging angle are

$$\varphi_k=\sum_{i=1}^{N}w_{k\varphi}(i)\varphi_k(i) \tag{28}$$

$$\theta_k = \sum_{i=1}^{N} w_{k\theta}(i)\theta_k(i) \tag{29}$$

3.3. DOA Tracking and Estimation Algorithm Based on Modified Particle Filtering

In the modified PF algorithm, the weights selection of the particles affects the filtering performance. When the particles are near the real state, the weights of the likehood function of the particles are large and *vice versa*. The particles with large weights are duplicated in the resampling process. Hence, from Equation (16), it is known that the selection of the likehood function is crucial for the modified PF algorithm.

The spectrum estimation of the MUSIC algorithm [8] is represented by:

$$P_{MUSIC}(\theta,\varphi) = \frac{1}{a^H(\theta,\varphi)\mathbf{GG}^H a(\theta,\varphi)} \tag{30}$$

Then, Equation (30) is used to improve the likehood observation function and the sampling particles are evaluated by

$$\hat{p}\left(\mathbf{Y}_k \middle| x_k(i)\right) = \frac{1}{a^H\left(x_k(i)\right)\mathbf{GG}^H a\left(x_k(i)\right)} \tag{31}$$

In [19], the exponential weights factor r was added to the likehood function, which increased the distinctiveness of the weights and solved the problem of invalid sampling of the particles, due to the flat of the likehood function in the low SNR. Hence, the likehood function is rewritten as

$$p\left(\mathbf{Y}_k \middle| x_k(i)\right) = \left\{\hat{p}\left(\mathbf{Y}_k \middle| x_k(i)\right) / \max_{x_k^i} \hat{p}\left(\mathbf{Y}_k \middle| x_k(i)\right)\right\}^r \tag{32}$$

where $r \in R^+$, and its value is determined by experiments.

In summary, the steps of the DOA tracking and estimation algorithm based on MPF and MUSIC (MPF-MUSIC) are summarized as follows:

Step 1: Initiation. The initial angles are estimated by random method. That is, the initial orientation is unknown and the angles are sampled from the uniform distribution with the number N and the weight $1/N$ of the particles:

$$\theta_0(i) \sim U[0,\pi/2], \varphi_0(i) \sim U[0,\pi], \dot{\varphi}_0(i), \dot{\theta}_0(i) \sim N(\mu_0, \varepsilon_0^2) \quad i = 1,\cdots,N$$

where $U[e_1,e_2]$ indicates the uniform distribution in the interval $[e_1,e_2]$ and N(μ0,ε20) indicates Gaussian distribution with mean μ0 and variance ε20.

Step 2: At the k moment, the receiving matrix is calculated according to Equation (10) and the noise subspace is obtained.

Step 3: $\left(\theta_k(i), \hat{\varphi}_{k-1}(i)\right)$ and $\left(\varphi_k(i), \hat{\theta}_{k-1}(i)\right)$ are sampled according to Equations (20) and (21), respectively; the likehood functions $p\left(\mathbf{Y}_k \middle| \varphi_k(i), \hat{\theta}_{k-1}(i)\right)$ and $p\left(\mathbf{Y}_k \middle| \theta_k(i), \hat{\varphi}_{k-1}(i)\right)$ are calculated according to Equation (32); and the weights of the particles w∗kθ(i) and w∗kφ(i) are calculated according to Equations (24) and (25), respectively.

Step 4: The normalized weights are calculated according to Equations (26) and (27).

Step 5: The estimated values of the state are calculated according to Equations (28) and (29).

Step 6: Let $k = k + 1$ and repeat steps from 2 to 6. If $k > T$ where T is the observation length, the algorithm stops.

The differences of the steps of the DOA tracking and estimation algorithm based on PF and PF-MUSIC are merely in Step 3 to Step 5 of which in MPF-MUSIC, that is:

Step 3*: (θ_k (i), φ_k (i)) is sampled according to Equation (16), the likehood function $p\left(\mathbf{Y}_k \middle| \left(\theta_k(i), \varphi_k(i)\right)\right)$ is calculated according to Equation (32) and the weights of the particles w∗k(i) are calculated according to Equation (17).

Step 4*: The normalized weights w_k (i) are calculated according to Equation (18) and resampled.

Step 5*: The estimated values of the state are calculated according to Equation (19).

4. SIMULATION AND ANALYSIS

Single Array Element Experiment

Experiment 1: dynamic DOA tracking and estimation experiment based on PF and the corresponding modified algorithms

Experimental conditions: single array element, $M = 1$, water is the propagation medium, the initial angle of the moving acoustic source is (30°, 20°) the moving interval is 1°, the snapshot is 256. In PF algorithm, the particle number is $N = 200$. In the modified PF algorithm, the particle number is $N = 100$, μ0=π/180,ε20=5×10−4, r = 13 which is determined by experiment. Let SNR = 5dB, observing the tracking deviation of the pitch angle and the azimuth angle, the experimental results is shown in Figure 1.

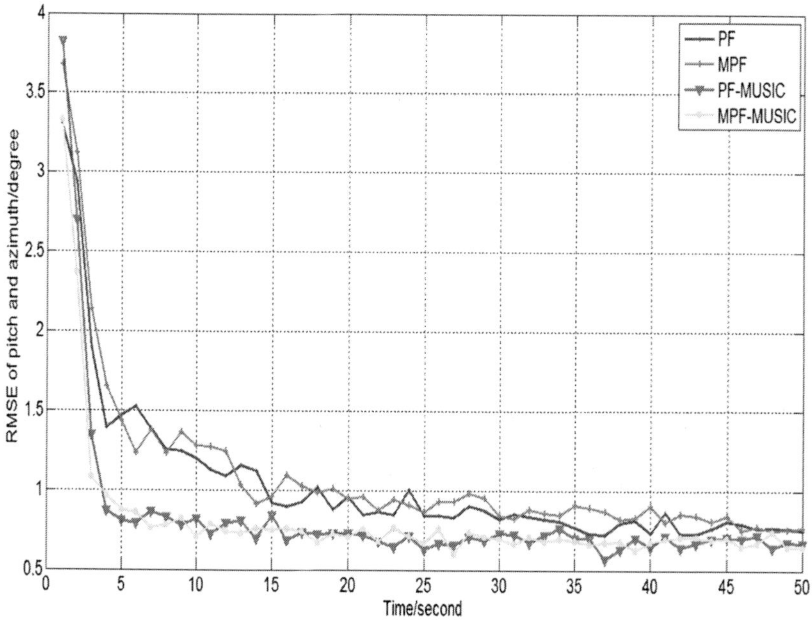

Figure 1. The RMSEs of the dynamic DOA tracking for the PF and the corresponding modified algorithms in different time.

As indicated in Figure 1, each algorithm accomplishes the angle tracking when the observation time is longer than 5 s. In the initial stage, the estimation deviation is large and the range of the deviation is within 4°. Since the initial of the azimuth angle is selected randomly from the uniform distribution, the deviation of the PF algorithm is large at the initial moment and tends to stabilize after seconds, as shown in Figure 1. After a period of time estimation and correction, the fluctuation range of the tracking deviation of each algorithm decreases and tends to stabilize. Usually, the samples of the initial moment can be determined by DOA estimation algorithm, which decreases the fluctuation error of the algorithm in the initial stage. Moreover, the dynamic RMSE of the MPF-MUSIC algorithm seems identical with that of the PF-MUSIC algorithm in different time, but the probability of convergence (PROC) is better than that of the PF-MUSIC algorithm, which will be discussed in Experiment 3.

Experiment 2: the experiment of the weight factor affecting the DOA tracking deviation based on standard PF filtering.

Experimental conditions: the values of the SNR are −5 dB, 0 dB and 5 dB, respectively. The simulation times of Monte Carlo experiments are 100. Other experimental parameters are same the as Experiment 1.

The Monte Carlo experiments are simulated independently for each exponential factor with the value of *r* changing from 1 to 22. The evaluation way of the estimation performance is the joint RMSE represented by:

$$\mathrm{RMSE}=\frac{1}{T}\sum_{j=1}^{T}\left\langle\sqrt{\frac{1}{100}\sum_{i=1}^{100}[(\hat{\theta}_{ij}-\theta_{ij})^2+(\hat{\varphi}_{ij}-\varphi_{ij})^2]}\right\rangle \tag{33}$$

where θˆij and φ^ij are the estimation values of the pitch angle and the azimuth angle in the *i*th simulation of the *j*th step. The simulation results are shown in Figure 2.

As indicated in Figure 2, the value of the RMSE decreases with the increase of *r*. When the value of *r* is large than 13, the tracking error of the algorithm tends to stabilize. Hence, r = 13 is selected in the simulation. Similarly, r = 6 is selected in the PF-MUSIC algorithm and MPF-MUSIC algorithm.

Experiment 3: the PROC experiment of the PF tracking algorithm under different SNR.

The definition of convergence:

$$\Upsilon_{ki}=\begin{cases}1 & |\hat{\theta}_i-\theta_i|+|\hat{\varphi}_i-\varphi_i|<\zeta \\ 0 & others\end{cases} \tag{34}$$

where θˆi and φ^i are the estimated value of the pitch angle and the azimuth angle in the *i*th simulation at the *k* moment, ζ is the threshold of the angle error and equal to 4 in the experiments. If the error is smaller than ζ, the estimated value is regarded as a successful estimation.

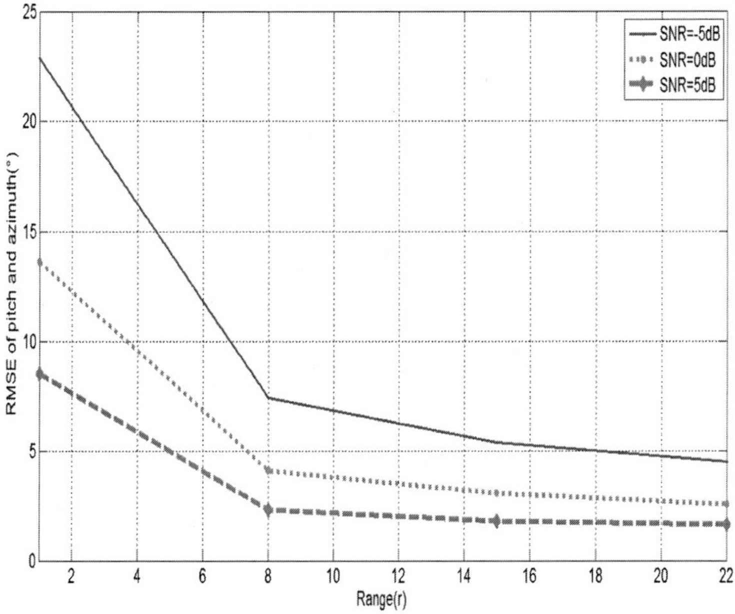

Figure 2. The RMSE sunder different exponential factors.

The calculation formula of the probability of convergence is:

$$PROC = \sum_{i=0}^{MC} \Upsilon_{ki} \Big/ MC \tag{35}$$

where MC is the times of the Monte Carlo experiments.

Experimental conditions: the range of the SNR is from −10 dB to 10 dB and the step size is 2. Other experimental parameters are the same as those in Experiment 1. The simulation times of Monte Carlo experiments are 500, the evaluation way of the performance is according to Equation (35) and the simulation results are shown in Figure 3.

As indicated in Figure 3, the MPF algorithm has higher PROC than that of PF algorithm and achieves 99% when the SNR is equal to 10 dB. The PROC of the MPF-MUSIC algorithm is comparable with that of the PF-MUSIC algorithm. Hence, the improvement of independent evaluation for the pitch angle and the azimuth angle increases the PROC of the angle, which is significant in practical engineering.

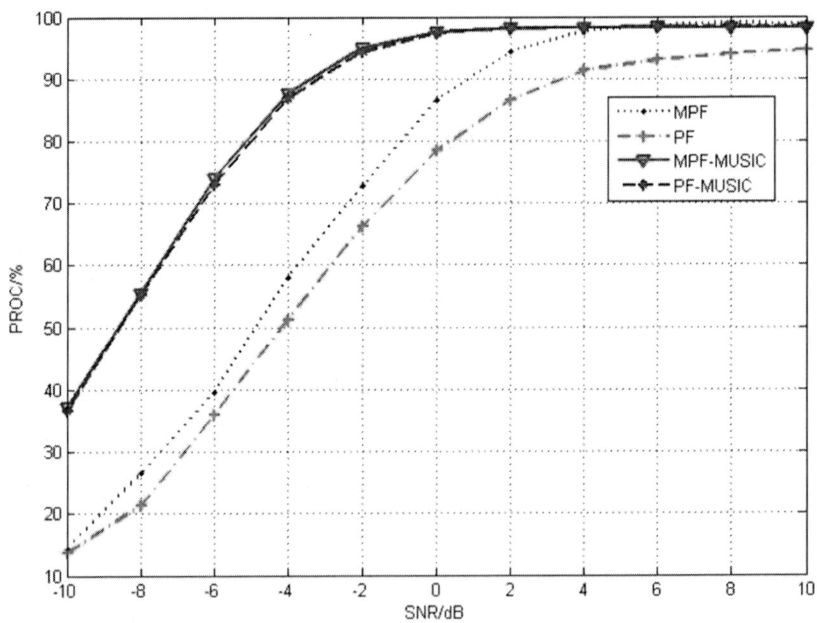

Figure 3. PROC with different SNRs.

Experiment 4: the DOA tracking performance experiments of PF tracking under different SNR.

Experimental conditions: the range of the SNR is from −10 dB to 10 dB and the step size is 5. Other experimental parameters are the same as those in Experiment 1. The simulation times of Monte Carlo experiments are 100, the evaluation way of the performance is according to Equation (33) and the simulation results are shown in Figure 4.

As indicated in Figure 4, the joint RMSE of each tracking algorithm decreases with the increase of the SNR. Compared with the standard PF algorithm, the MPF algorithm proposed in the article has better estimation performance. Among the different PF algorithms, the tracking error of the PF-MUSIC algorithm is smaller than that of the PF algorithm. When the SNR is more than −5 dB, the tracking error is within 2°. Moreover, the RMSEs of the MPF-MUSIC algorithm under different SNRs are almost identical with those of the PF-MUSIC algorithm, but the PROC is also better than that of the PF-MUSIC algorithm, as discussed above.

5. CONCLUSIONS

This article studies the target tracking algorithm based on acoustic vector sensor array and proposes a modified particle filtering algorithm based on sequential importance sampling. The proposed algorithm solves the problems of large deviation of either pitch angle or azimuth angle leading to the other angle that cannot be duplicated and propagated through resampling in the two dimensional angle estimation. Furthermore, the MUSIC spectrum estimation algorithms are introduced as the observation likehood function of the modified particle filtering algorithm to improve the tracking performance and the combination of the modified particle filtering algorithm and MPF-MUSIC algorithm is proposed. As indicated in the experiments, the performance of MPF algorithm is better than that of PF algorithm, which has higher PROC. Furthermore, the performance of PF-MUSIC and MPF-MUSIC algorithms are better than that of PF and MPF algorithms in the aspects of RMSE, respectively. However, the MPF-

MUSIC algorithm achieves comparable RMSEs and PROC with those of the PF-MUSIC algorithm.

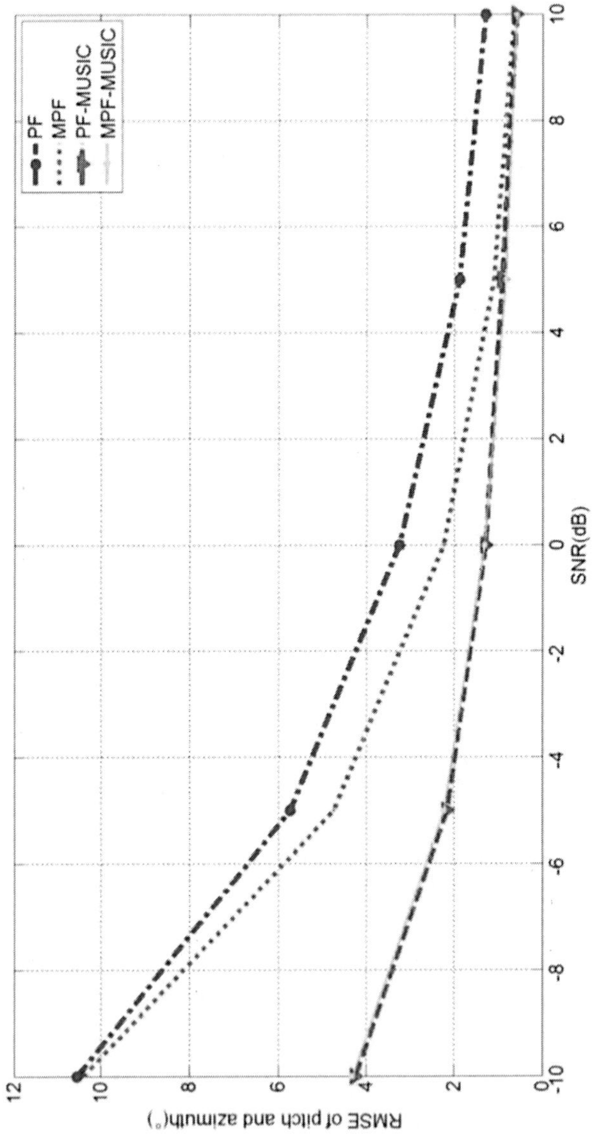

Figure 4. The RMSEs under different SNRs.

ACKNOWLEDGMENTS

This research is supported by the National Natural Science Foundation of China (Grant No. 51475198), China Postdoctoral Science Foundation (Grant No. 2015M570270), Jilin Province Youth Science Foundation (Grant No. 20140520064JH) and Jilin Province Education Department Project (Grant No. 2014LY502L20).

AUTHOR CONTRIBUTIONS

For this research, Xinbo Li derived the algorithm, proposed the model, designed the experiments and drafted the manuscript. Haixin Sun and Liangxu Jiang collected the data, and validated the results of the experiments. Yaowu Shi and Yue Wu checked the manuscript.

CONFLICTS OF INTEREST

The authors declare no conflict of interest.

REFERENCES

1. Wu, Y.I.; Wong, K.T.; Lau, S.K. The Acoustic Vector-Sensor's Near-Field Array-Manifold. *IEEE Trans. Signal Process.* **2010**, *58*, 3946–3951.

2. Yang, S.; Wong, K.T. Three-Dimensional Localization of a Near-Field Emitter of Unknown Spectrum, using an Acoustic Vector Sensor Corrupted by Additive Noise of Unknown Spectrum. *IEEE Trans. Aerosp. Electron. Syst.* **2013**,*49*, 1035–1041.

3. Wong, K.T. Acoustic Vector-Sensor FFH "Blind" Beamforming & Geolocation. *IEEE Trans. Aerosp. Electron. Syst.* **2010**,*46*, 444–448.

4. Ping, K.T.; Wong, K.T.; Song, Y. An Hybrid Cramér-Rao Bound in Closed Form for Direction-of-Arrival Estimation by an "Acoustic Vector Sensor" with Gain-Phase Uncertainties. *IEEE Trans. Signal Process.* **2014**, *62*, 2504–2516.

5. Tam, P.K.; Wong, K.T. Cramér-Rao Bounds for Direction Finding by an Acoustic Vector Sensor under Nonideal Gain-Phase Responses, Noncollocation, or Nonorthogonal Orientation. *IEEE Sens. J.* **2009**, *9*, 969–982.

6. Yang, S.; Wong, K.T. Azimuth-elevation direction finding using a microphone and three orthogonal velocity sensors as a non-collocated subarray. *J. Acoust. Soc. Am.* **2013**, *133*, 1987–1995.

7. Yang, S.; Wong, K.T. Acoustic direction finding using a spatially spread tri-axial velocity sensor. *IEEE Trans. Aerosp. Electron. Syst.* **2015**, *51*, 834–842.

8. Nehorai, A.; Paldi, E. Acoustic vector sensor array processing. *IEEE Trans. Signal Process.* **1994**, *42*, 2481–2491.

9. Hawkes, M.; Nehorai, A. Acoustic vector-sensor beamforming and Capon direction estimation. *IEEE Trans. Signal Process.* **1998**, *46*, 2291–2304.

10. Tichavsky, P.; Wong, K.T.; Zoltowski, M.D. Near-field/far-field azimuth and elevation angle estimation using a single vector hydrophone. *IEEE Trans. Signal Process.* **2001**, *49*, 2498–2510.

11. Chen, G.C.; Hodgkiss, W.S. VLF source localization with a freely drifting acoustic sensor array. *IEEE J. Ocean. Eng.* **1993**, *18*, 209–223.

12. Ratnarajah, T. An H∞ approach to multi-source tracking. In Proceedings of the 1998 IEEE International Conference on Acoustics, Speech, and Signal Processing, Seattle, WA, USA, 15 May 1998; pp. 2205–2208.

13. Liu, X.; Xiang, J.; Zhou, Y. Passive tracking and size estimation of volume target based on acoustic vector intensity.*Chin. J. Acoust.* **2001**, *20*, 225–238.

14. Awad, M.K.; Wong, K.T. Recursive Least-Squares Source Tracking using One Acoustic Vector Sensor. *IEEE Trans. Aerosp. Electron. Syst.* **2012**, *48*, 3073–3083.

15. Awad, M.K.; Wong, K.T. Errata: Recursive least-squares source tracking using one acoustic vector sensor. *IEEE Trans. Aerosp. Electron. Syst.* **2013**, *49*, 688–692.

16. Zhong, X.; Premkumar, A.B. Particle Filtering Approaches for Multiple Acoustic Source Detection and 2-D Direction of Arrival Estimation Using a Single Acoustic Vector Sensor. *IEEE Trans. Signal Process.* **2012**, *60*, 4719–4733.

17. Zhong, X.H.; Premkumar, A.B.; Madhukun, A.S. Particle filtering and posterior Cramér-Rao Bound for 2-D direction of arrival tracking using an acoustic vector sensor. *IEEE Sens. J.* **2012**, *12*, 363–377.

18. Zhong, X.H.; Hari, V.N.; Premkumar, A.B. Particle filtering with enhanced likelihood model for underwater acoustic source DOA tracking. In Proceedings of the Oceans'11 IEEE Santander Conference, Santander, Spain, 6–9 June 2011; pp. 1–6.

19. Zhong, X.; Premkumar, A.B.; Wang, W. Direction of arrival tracking of an underwater acoustic source using particle filtering: Real data experiments. In Proceedings of the IEEE Tencon Spring Conference, Xi'an, China, 22–25 October 2013; pp. 420–424.

20. Zhong, X.H.; Premkumar, A.B.; Madhukumar, A.S. Direction of arrival tracking in impulsive noise using particle filtering with fractional lower order moment likelihood. In Proceedings of the International Conference on Information, Communications and Signal Processing, Singapore, 13–16 December 2011; pp. 1–5.

21. Gordon, N.; Salmond, D.; Smith, A.F. Novel approach to nonlinear/non-Gaussian Bayesian state estimation. *Proc. Inst. Elect. Eng.* **1993**, *140*, 107–113.

22. Liu, J.S.; Chen, R. Sequential Monte Carlo Methods for Dynamic Systems. *J. Am. Stat. Assoc.* **1998**, *93*, 1032–1044.

23. Arulampalam, M.; Maskell, S.; Gordon, N.; Clapp, T. A tutorial on particle filters for online nonlinear/non-Gaussian Bayesian tracking. *IEEE Trans. Signal Process.* **2002,** *50*, 174–188.

24. Ristic, B.; Arulampalam, S.; Gordon, N. *Beyond the Kalman Filter: Particle Filters for Tracking Applications*; Artech House: London, UK, 2004.

25. Wiese, T.; Claussen, H.; Rosca, J. Particle filter based DOA estimation for multiple source tracking (MUST). In Proceedings of the 11th Asilomar Conference on Circuits Systems & Computers, California, CA, USA, 6–9 November 2011; pp. 624–628.

26. Djurić, P.M.; Venula, M.; Bugallo, M.F. Target Tracking by Particle Filtering in Binary Sensor Networks. *IEEE Trans. Signal Process.* **2008**, *56*, 2229–2238.

27. Weber, R.J.; Huang, Y.; Weber, R.J. Analysis for Capon and MUSIC DOA estimation algorithms. In Proceedings of the IEEE Antennas and Propagation Society International Symposium, Cgarkeston, SC, USA, 1–5 June 2009; pp. 1–4.

28. Wong, K.T.; Zoltowski, M.D. Self-initiating MUSIC-based direction finding in underwater acoustic particle velocity-field beamspace. *IEEE J. Ocean. Eng.* **2000**, *25*, 262–273.

29. Wong, K.T.; Zoltowski, M.D. Root-MUSIC-based azimuth-elevation angle-of-arrival estimation with uniformly spaced but arbitrarily oriented velocity hydrophones. *IEEE Trans. Signal Process.* **1999**, *47*, 3250–3260.

30. Zoltowski, M.D; Wong, K.T. Closed-form eigenstructure-based direction finding using arbitrary but identical subarrays on a sparse uniform Cartesian array grid. *IEEE Trans. Signal Process* **2000**, *48*, 2205–2210.

CHAPTER 10

Robust Hammerstein Adaptive Filtering under Maximum Correntropy Criterion

Zongze Wu [1], Siyuan Peng [1], Badong Chen [2,*] and Haiquan Zhao [3]

[1] School of Electronic and Information Engineering, South China University of Technology, Guangzhou 510640, China

[2] School of Electronic and Information Engineering, Xi'an Jiaotong University, Xi'an 710049, China

[3] School of Electrical Engineering, Southwest Jiaotong University, Chengdu 610031, China

ABSTRACT

The maximum correntropy criterion (MCC) has recently been successfully applied to adaptive filtering. Adaptive algorithms under MCC show strong robustness against large outliers. In this work, we apply the MCC criterion to develop a robust Hammerstein adaptive filter. Compared with the traditional Hammerstein adaptive filters, which are usually derived based on the well-known mean square error (MSE) criterion, the proposed algorithm can achieve better convergence performance especially in the presence of impulsive non-Gaussian (e.g., α-stable) noises. Additionally, some theoretical results concerning the convergence behavior are also obtained. Simulation examples are presented to confirm the superior performance of the new algorithm.

KEYWORDS

Hammerstein adaptive filtering; MCC; nonlinear system identification

1. INTRODUCTION

Nonlinear system identification is still an active research area [1]. Although linear systems have established a solid theory [2], most practical systems (e.g., hands-free telephone systems) may be more adequately represented as a nonlinear model. One of the main challenges for nonlinear system identification is the choice of an appropriate nonlinear filtering structure that accurately captures the characteristics of the underlying nonlinear system. A common structure used in nonlinear modeling is the *block-oriented* representation. The Wiener model and the Hammerstein model are two typical block-oriented nonlinear models [3]. Specifically, the Wiener model consists of a cascade of a linear time invariant (LTI) filter followed by a static nonlinear function, indicated as a linear-nonlinear (LN) model [4,5,6], and the Hammerstein model consists of a cascade of a static nonlinear function follow by a LTI filter, known as a nonlinear-linear (NL) model [7,8,9,10,11,12,13,14, 15,16,17, 18,19]. Other nonlinear models include neural networks (NNs) [20], Volterra adaptive filters (VAFs) [21], kernel adaptive filters (KAF) [22,23,24,25], among others.

Hammerstein filters can accurately model many real-world systems and, as a consequence, they have been successfully used in various applications of engineering [26,27,28,29]. Due to its simplicity and efficiency, the mean square error (MSE) criterion has been widely applied in Hammerstein adaptive filtering [30]. Adaptive algorithms under MSE usually perform very well when the desired signals are disturbed by Gaussian noises. However, when the desired signals are disturbed by non-Gaussian noises, especially in the presence of large outliers (observations that significantly deviate from the bulk of data), the performance of the MSE based algorithms may deteriorate rapidly. Actually, MSE is rather sensitive to outliers. In most practical situations, heavy-tailed impulsive noises may occur, which often cause large outliers. For instance, different types of artificial noises in electronic devices, atmospheric noises, and lighting spikes in natural phenomena, can be described as an impulsive noise [31,32].

In this work, instead of using the MSE criterion, we apply the maximum correntropy criterion (MCC) to develop a robust Hammerstein adaptive filtering algorithm. Correntropy is a nonlinear similarity measure between two signals [33,34]. The MCC aims at maximizing the similarity (measured by correntropy) between the model output and the desired response such that the adaptive model is as close as possible to the unknown system. It has been shown that, the MCC in terms of the stability and accuracy, is very robust with respect to impulsive noises [33,34,35,36,37,38,39]. Compared with the traditional Hammerstein adaptive filtering algorithms based on the MSE criterion, the new algorithm can achieve better performance especially in the presence of impulsive non-Gaussian noises.

The organization of the rest of the paper is as follows. In Section 2, after briefly introducing the correntropy, we derive a Hammerstein adaptive filtering algorithm under MCC criterion. In Section 3, we carry out the convergence analysis. InSection 4, we present simulation examples to demonstrate the superior performance of the proposed algorithm. Finally, we give the conclusion in Section 5.

2. HAMMERSTEIN ADAPTIVE FILTERING UNDER THE MAXIMUM CORRENTROPY CRITERION

Figure 1 shows the structure of a Hammerstein adaptive filter under MCC criterion, where the filter consists of a polynomial memoryless nonlinearity followed by a linear FIR filter. This structure has been commonly used in Hammerstein adaptive filtering [8,9,27]. As shown in Figure 1, under the MCC criterion, the parameters of the linear and nonlinear parts are adjusted to maximize the correntropy between the model output and desired response.

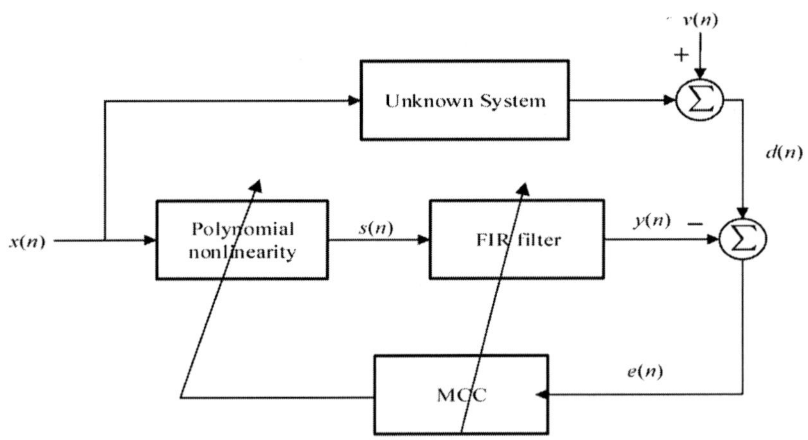

Figure 1. Structure of a Hammerstein adaptive filter under maximum correntropy criterion (MCC) criterion.

2.1. Correntropy

Correntropy is a nonlinear similarity measure between two signals. Given two random variables X and Y, the correntropy is [33,34, 35,36,37,38,39]

$$V(X,Y)=E[\kappa(X,Y)]=\int \kappa(x,y)f_{XY}(x,y)dxdy \tag{1}$$

where $E[\cdot]$ denotes the expectation operator, $\kappa(\cdot,\cdot)$ is a *shift-invariant* Mercer kernel, and $f_{XY}(x, y)$ stands for the probability density function (PDF) of (X, Y). The most widely used kernel in correntropy is the Gaussian kernel, given by

$$\kappa_\sigma(x,y)=\frac{1}{\sigma\sqrt{2\pi}}\exp(-\frac{e^2}{2\sigma^2}) \tag{2}$$

where $e = x - y$, and σ stands for the kernel bandwidth. In this work, without being mentioned otherwise, the kernel function is a Gaussian kernel. In practical situations, the join distribution of X and Y is usually unknown and only a finite number of data $\{(d(i),y(i))\}_{i=1}^{K}$ are available. In these cases, one can use a sample mean estimator of the correntropy:

$$\hat{V}_{N,\sigma}(X,Y)=\frac{1}{K}\sum_{i=1}^{K}\kappa_\sigma(d(i)-y(i)) \tag{3}$$

The optimization cost under MCC is thus

$$\max J_{MCC}=\sum_{i=1}^{K}\kappa_\sigma(e(i)) \tag{4}$$

where $e(i) = d(i) - y(i)$. We can evaluate the sensitivity (derivative) of the MCC cost J_{MCC} with respect to the error $e(i)$,

$$\frac{\partial J_{MCC}}{\partial e(i)}=\frac{-1}{\sqrt{2\pi}\sigma^3}\cdot\exp\left(\frac{-e^2(i)}{2\sigma^2}\right)e(i) \tag{5}$$

The derivative curves of $-J_{MCC}$ for different kernel widths are illustrated in Figure 2. As one can see, when the magnitude of error is very large, the derivative will become rather small especially for a smaller kernel width. Therefore, the MCC training is insensitive (hence robust) to a large error.

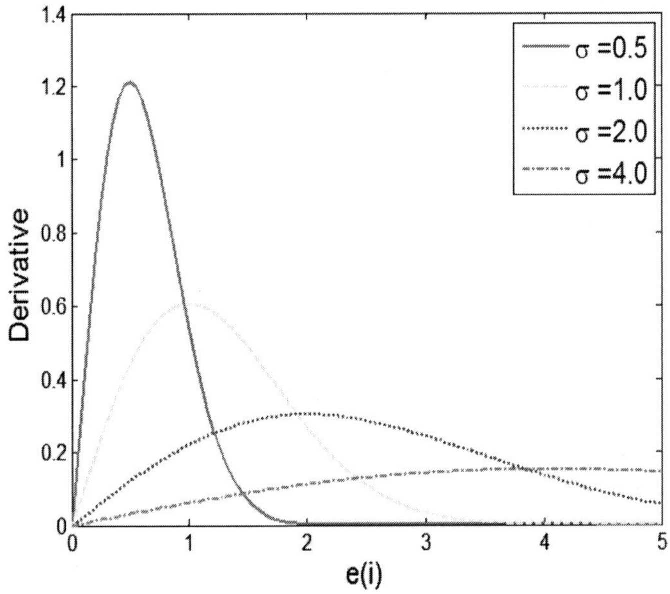

Figure 2. Derivative curves of $-J_{MCC}$ with respect to $e(i)$ for different kernel widths.

2.2. Hammerstein Adaptive Filtering

Assuming that the input-output mapping of the memoryless polynomial nonlinearity is

$$s(n) = p_1 x(n) + p_2 x^2(n) + \cdots + p_M x^M(n) \tag{6}$$

where M and p_M denote the polynomial order and the m-th order coefficient, Expression (6) can be rewritten as

$$s(n) = \mathbf{p}^T(n)\mathbf{x}_p(n) \tag{7}$$

where $\mathbf{x}_p(n) = [x(n)\ x^2(n)\cdots x^M(n)]^T$ is the polynomial regressor, and $\mathbf{p}(n) = [p_1\ p_2\cdots p_M]^T$ is the polynomial coefficient vector. The output of the FIR filter can be expressed as

$$y(n) = \mathbf{w}^T(n)\mathbf{s}(n) \tag{8}$$

where $\mathbf{w}(n) = [w_0\ w_1\cdots w_{N-1}]^T$ is the FIR weight vector, and $\mathbf{s}(n) = [s(n)\ s(n-1)\cdots s(n-N+1)]^T$ is the FIR input vector, with Nbeing the FIR memory size. Let $\mathrm{X}(n) = [\mathbf{x}_p(n)\ \mathbf{x}_p(n-1)\cdots \mathbf{x}_p(n-N+1)]^T$. Then we have

$$\mathbf{s}(n) = \mathrm{X}^T(n)\mathbf{p}(n) \tag{9}$$

Combining Equations (8) and (9) yields

$$y(n) = \mathbf{w}^T(n)\mathbf{s}(n) = \mathbf{w}^T(n)\mathrm{X}^T(n)\mathbf{p}(n) \tag{10}$$

Assume that the unknown system that needs to be identified is also a Hammerstein system with parameter vectors p∗=[p∗1 p∗2···p∗M]T and w∗=[w∗0 w∗1···w∗N−1]T. Then, the desired signal can be expressed as

$$d(n) = \mathbf{w}^{*T}\mathrm{X}^T(n)\mathbf{p}^* + v(n) \tag{11}$$

where $v(n)$ stands for an additive disturbance noise. The error signal can then be calculated as $e(n) = d(n) - \mathbf{w}^T(n)\mathrm{X}^T(n)\mathbf{p}(n)$. In the following, we derive an adaptive algorithm to estimate the Hammerstein parameter vectors using MCC instead of MSE as an optimization criterion. Let us consider the following cost function

$$\begin{aligned} J_{MCC}(\mathbf{p},\mathbf{w}) &= \sum_{j=n-L+1}^{n} \kappa_{\sigma}(d(j), y(j)) \\ &= \frac{1}{\sqrt{2\pi}\sigma} \sum_{j=n-L+1}^{n} \exp\left(\frac{-e^2(j)}{2\sigma^2}\right) \end{aligned} \tag{12}$$

where $e(j) = d(j) - y(j)$, and L denotes the sliding data length. Then, a steepest ascent algorithm for estimating the polynomial coefficient vector can be derived as follows:

$$\begin{aligned} \frac{\partial J_{MCC}(\mathbf{p},\mathbf{w})}{\partial \mathbf{p}(n)} &= \frac{1}{\sqrt{2\pi}\sigma^3} \cdot \sum_{j=n-L+1}^{n} \exp\left(\frac{-e^2(j)}{2\sigma^2}\right) e(j) \cdot \frac{\partial e(j)}{\partial \mathbf{p}(j)} \\ &= \frac{1}{\sqrt{2\pi}\sigma^3} \cdot \sum_{j=n-L+1}^{n} \exp\left(\frac{-e^2(j)}{2\sigma^2}\right) e(j) \cdot \mathrm{X}(j)\mathbf{w}(j) \end{aligned} \tag{13}$$

$$\mathbf{p}(n+1) = \mathbf{p}(n) + \mu_p \frac{\partial J_{MCC}(\mathbf{p},\mathbf{w})}{\partial \mathbf{p}(n)} = \mathbf{p}(n) + \frac{\mu_p}{\sqrt{2\pi}\sigma^3} \cdot \sum_{j=n-L+1}^{n} \exp\left(\frac{-e^2(j)}{2\sigma^2}\right) e(j) \cdot \mathrm{X}(j)\mathbf{w}(j) \tag{14}$$

In a similar way, we propose the following weight update equation for the coefficients of the FIR filter:

$$\begin{aligned} \frac{\partial J_{MCC}(\mathbf{p},\mathbf{w})}{\partial \mathbf{w}(n)} &= \frac{1}{\sqrt{2\pi}\sigma^3} \cdot \sum_{j=n-L+1}^{n} \exp\left(\frac{-e^2(j)}{2\sigma^2}\right) e(j) \cdot \frac{\partial e(j)}{\partial \mathbf{w}(j)} \\ &= \frac{1}{\sqrt{2\pi}\sigma^3} \cdot \sum_{j=n-L+1}^{n} \exp\left(\frac{-e^2(j)}{2\sigma^2}\right) e(j) \cdot \mathrm{X}^T(j)\mathbf{p}(j) \end{aligned} \tag{15}$$

$$\mathbf{w}(n+1) = \mathbf{w}(n) + \mu_w \frac{\partial J_{MCC}(\mathbf{p},\mathbf{w})}{\partial \mathbf{w}(n)} = \mathbf{w}(n) + \frac{\mu_w}{\sqrt{2\pi}\sigma^3} \cdot \sum_{j=n-L+1}^{n} \exp\left(\frac{-e^2(j)}{2\sigma^2}\right) e(j) \cdot \mathrm{X}^T(j)\mathbf{p}(j) \tag{16}$$

In Equations (14) and (16), μ_p and μ_w are, respectively, step-sizes for polynomial nonlinearity subsystem and FIR subsystem. In this work, for simplicity we consider only the stochastic gradient based algorithm (*i.e.*, $L = 1$). In this case, we have

$$\mathbf{p}(n+1) = \mathbf{p}(n) + \eta_p \cdot \exp\left(\frac{-e^2(n)}{2\sigma^2}\right) e(n) \cdot \mathbf{X}(n)\mathbf{w}(n) \tag{17}$$

$$\mathbf{w}(n+1) = \mathbf{w}(n) + \eta_w \cdot \exp\left(\frac{-e^2(n)}{2\sigma^2}\right) e(n) \cdot \mathbf{X}^T(n)\mathbf{p}(n) \tag{18}$$

where $\eta_p = \frac{\mu_p}{\sqrt{2\pi\sigma^3}}$, and $\eta_w = \frac{\mu_w}{\sqrt{2\pi\sigma^3}}$. The above update equations are referred to as the Hammerstein adaptive filtering algorithm under MCC criterion, whose pseudocodes are presented in Algorithm 1. The proposed algorithm is in form similar to the traditional Hammerstein adaptive filters under MSE criterion [7], but the step-sizes are different.

Algorithm 1: Hammerstein adaptive filtering Algorithm under MCC.

Parameters setting: μ_p, μ_w, σ

Initialization: $\mathbf{p}(0)$, $\mathbf{w}(0)$

For $n = 1, 2, \ldots$ **do**

(1) $\mathbf{s}(n) = \mathbf{X}^T(n)\mathbf{p}(n)$

(2) $y(n) = \mathbf{w}^T(n)\mathbf{s}(n)$

(3) $e(n) = d(n) - y(n)$

(4) $\mathbf{p}(n+1) = \mathbf{p}(n) + \eta_p \cdot \exp\left(\frac{-e^2(n)}{2\sigma^2}\right) e(n) \cdot \mathbf{X}(n)\mathbf{w}(n)$

(5) $\mathbf{w}(n+1) = \mathbf{w}(n) + \eta_w \cdot \exp\left(\frac{-e^2(n)}{2\sigma^2}\right) e(n) \cdot \mathbf{X}^T(n)\mathbf{p}(n)$

End for

---

3. CONVERGENCE ANALYSIS

3.1. Stability Analysis

Using the Taylor series expansion of the error $e(n+1)$ around the instant n and keeping only the linear term, we have [4,7,40]

$$e(n+1)=e(n)+\frac{\partial e(n)}{\partial \mathrm{p}(n)}\bigg|_{\mathrm{w}(n)=const}\Delta\mathrm{p}(n)+\frac{\partial e(n)}{\partial \mathrm{w}(n)}\bigg|_{\mathrm{p}(n)=const}\Delta\mathrm{w}(n)+h.o.t \tag{19}$$

where $h.o.t$ denotes higher-order terms. Combining Equations (11), (17) and (18), we can obtain

$$\frac{\partial e(n)}{\partial \mathrm{p}(n)}=-\mathrm{X}(n)\mathrm{w}(n) \tag{20}$$

$$\frac{\partial e(n)}{\partial \mathrm{w}(n)}=-\mathrm{X}^T(n)\mathrm{p}(n) \tag{21}$$

$$\Delta\mathrm{p}(n)=\eta_p\cdot\exp\left(\frac{-e^2(n)}{2\sigma^2}\right)e(n)\cdot\mathrm{X}(n)\mathrm{w}(n) \tag{22}$$

$$\Delta\mathrm{w}(n)=\eta_w\cdot\exp\left(\frac{-e^2(n)}{2\sigma^2}\right)e(n)\cdot\mathrm{X}^T(n)\mathrm{p}(n) \tag{23}$$

Substituting Equations (20)–(23) in Equation (19), and after simple manipulation, we have

$$e(n+1)=\left[1-\eta_p\cdot\exp\left(\frac{-e^2(n)}{2\sigma^2}\right)\cdot\left\|\mathrm{X}(n)\mathrm{w}(n)\right\|^2-\eta_w\cdot\exp\left(\frac{-e^2(n)}{2\sigma^2}\right)\cdot\left\|\mathrm{X}^T(n)\mathrm{p}(n)\right\|^2\right]e(n) \tag{24}$$

To ensure the stability of the proposed algorithm, we must assure that $|e(n+1)| \leq |e(n)|$, and hence

$$\left| 1-\eta_p \cdot \exp\left(\frac{-e^2(n)}{2\sigma^2}\right) \cdot \left\| \mathrm{X}(n)\mathrm{w}(n) \right\|^2 - \eta_w \cdot \exp\left(\frac{-e^2(n)}{2\sigma^2}\right) \cdot \left\| \mathrm{X}^T(n)\mathrm{p}(n) \right\|^2 \right| \leq 1 \tag{25}$$

which yields

$$0 < \eta_p \cdot \left\| \mathrm{X}(n)\mathrm{w}(n) \right\|^2 + \eta_w \cdot \left\| \mathrm{X}^T(n)\mathrm{p}(n) \right\|^2 \leq \frac{2}{\exp\left(\frac{-e^2(n)}{2\sigma^2}\right)} \tag{26}$$

Since $\exp\left(\frac{-e^2(n)}{2\sigma^2}\right) \leq 1$, the following condition guarantees convergence:

$$0 < \eta_p \cdot \left\| \mathrm{X}(n)\mathrm{w}(n) \right\|^2 + \eta_w \cdot \left\| \mathrm{X}^T(n)\mathrm{p}(n) \right\|^2 \leq 2 \tag{27}$$

Remark 1. The derived bound on step-sizes is only of theoretical importance as in general, Equation (27) cannot be verified in a practical situation. Similar theoretical results can be found in [7].

3.2. Steady-State Mean Square Performance

We denote $e_{pw}(n)$ the *a priori* error of the whole system, $e_p(n)$ the *a priori* error when only the nonlinear part is adapted while the linear filter is fixed, and $e_w(n)$ the *a priori* error when only the linear filter is adapted while the nonlinear part is fixed. Let $\mathrm{H}_p = \lim_{n\to\infty} E\left[e_p^2(n)\right]$, $\mathrm{H}_w = \lim_{n\to\infty} E\left[e_w^2(n)\right]$, and $\mathrm{H}_{pw} = \lim_{n\to\infty} E\left[e_{pw}^2(n)\right]$ be the steady-state *excess mean square errors* (EMSEs). In addition, we denote

$$f(e(i)) = \exp\left(\frac{-e^2(i)}{2\sigma^2}\right)e(i) \tag{28}$$

Before evaluating the theoretical values of the steady-state EMSEs, we make the following assumptions:

A. The noise $v(n)$ is zero-mean, independent, identically distributed, and is independent of the input X(n), sˆ(n) and$e(n)$.

B. The *a priori* errors $e_p(n)$ and $e_w(n)$ are zero-mean Gaussian, and independent of the noise $v(n)$.

C. $||\mathrm{X}(n)\mathbf{w}(n)||^2$ and $||\mathrm{X}^T(n)\mathbf{p}(n)||^2$ are asymptotically uncorrelated with $f^2(e(n))$, that is

$$\lim_{n\to\infty} E\left[\left\|\mathrm{X}(n)\mathrm{w}(n)\right\|^2 f^2(e(n))\right] = Tr(R_{\overline{W}X}) \lim_{n\to\infty} E\left[f^2(e(n))\right] \tag{29}$$

$$\lim_{n\to\infty} E\left[\left\|\mathrm{X}^T(n)\mathrm{p}(n)\right\|^2 f^2(e(n))\right] = Tr(R_{XP}) \lim_{n\to\infty} E\left[f^2(e(n))\right] \tag{30}$$

where $R_{\overline{W}X} = E\left[(\mathrm{X}(n)\mathrm{w}(n))(\mathrm{X}(n)\mathrm{w}(n))^T\right]$ and $R_{XP} = E\left[(\mathrm{X}^T(n)\mathrm{p}(n))(\mathrm{X}^T(n)\mathrm{p}(n))^T\right]$ are the covariance matrices, and Tr(·) denotes the trace operator.

Remark 2. For the assumption (A), it is very common to assume that the noise is independent of the regression vector [41,42,43]. In addition, the noise is often restricted to be zero-mean, identically distributed [33,34,35]. As discussed in [44,45], the assumption (B) is reasonable for long adaptive filters. Since $e_p(n)$ is the *a priori error* when only the nonlinear part is adapted while the linear filter is fixed, we have the approximation $\mathbf{w}^* \approx \mathbf{w}(n)$ such that $\mathbf{w}(n)$ is asymptotically uncorrelated with $f^2(e(n))$. Due to the independent assumption (A), X(n) is also asymptotically uncorrelated with $f^2(e(n))$. So $||\mathrm{X}^T(n)\mathbf{w}(n)||^2$ is asymptotically uncorrelated with $f^2(e(n))$. Similarly, $||\mathrm{X}^T(n)\mathbf{p}(n)||^2$ is asymptotically uncorrelated with$f^2(e(n))$. Therefore, the assumption (C) is rational.

When only the polynomial part with parameter vector **p** is adapted, the error $e_p(n)$ is

$$e_p(n) = \mathrm{w}^{*T}\mathrm{X}^T(n)\mathrm{p}^* - \mathrm{w}^T(n)\mathrm{X}^T(n)\mathrm{p}(n) \approx \mathrm{w}^T(n)\mathrm{X}^T(n)\tilde{\mathrm{p}}(n) \qquad (31)$$

where p˜(n)=p∗−pT(n). In Equation (31), we use the approximation $\mathbf{w}^* \approx \mathbf{w}(n)$ at steady-state. From Equation (17), it follows easily that

$$\tilde{\mathrm{p}}(n+1) = \tilde{\mathrm{p}}(n) - \eta_p \cdot f(e(n)) \cdot \mathrm{X}(n)\mathrm{w}(n) \qquad (32)$$

Squaring both sides of Equation (32), we have

$$\left\|\tilde{\mathrm{p}}(n+1)\right\|^2 = \left\|\tilde{\mathrm{p}}(n)\right\|^2 - 2\eta_p \cdot f(e(n)) \cdot e_p(n) + \eta_p^2 \cdot f^2(e(n)) \cdot \left\|\mathrm{X}(n)\mathrm{w}(n)\right\|^2 \qquad (33)$$

Taking the expectations of the both sides of Equation (33) yields

$$E\left[\left\|\tilde{\mathrm{p}}(n+1)\right\|^2\right] = E\left[\left\|\tilde{\mathrm{p}}(n)\right\|^2\right] - 2\eta_p \cdot E\left[f(e(n)) \cdot e_p(n)\right] + \eta_p^2 \cdot E\left[f^2(e(n)) \cdot \left\|\mathrm{X}(n)\mathrm{w}(n)\right\|^2\right] \qquad (34)$$

Assuming the filter is stable and attains the steady state, it holds

$$\lim_{n\to\infty} E\left[\left\|\tilde{\mathrm{p}}(n+1)\right\|^2\right] = \lim_{n\to\infty} E\left[\left\|\tilde{\mathrm{p}}(n)\right\|^2\right] \qquad (35)$$

Combining Equations (34) and (35) and the above assumptions, we obtain

$$2 \cdot \lim_{n\to\infty} E\left[f(e(n)) \cdot e_p(n)\right] = \eta_p Tr(R_{\overline{W}X}) \lim_{n\to\infty} E\left[f^2(e(n))\right] \qquad (36)$$

In order to derive a theoretical value of the steady-state EMSE, we consider two cases below.

Case A. Gaussian Noise

Recalling that $e(n) = e_p(n) + v(n)$, and assuming that the noise $v(n)$ is zero-mean Gaussian, with variance $\varsigma 2v$, we get [34]

$$\lim_{n\to\infty} E\left[f(e(n))\cdot e_p(n)\right]=\frac{\sigma^3 \mathrm{H}_p}{(\sigma^2+\varsigma_v^2+\mathrm{H}_p)^{3/2}} \tag{37}$$

where σ_e^2 denotes the variance of the error, and $\sigma_e^2 = E[e_p^2(n)] + \varsigma_v^2$. Similarly, we obtain [29]

$$\lim_{n\to\infty} E\left[f^2(e(n))\right]=\frac{\sigma^3(\mathrm{H}_p+\varsigma_v^2)}{(\sigma^2+2\varsigma_v^2+2\mathrm{H}_p)^{3/2}} \tag{38}$$

Substituting Equations (37) and (38) into Equation (36), we have

$$2\cdot\frac{\sigma^3 \mathrm{H}_p}{(\sigma^2+\varsigma_v^2+\mathrm{H}_p)^{3/2}}=\eta_p Tr(R_{\overline{WX}})\frac{\sigma^3(\mathrm{H}_p+\varsigma_v^2)}{(\sigma^2+2\varsigma_v^2+2\mathrm{H}_p)^{3/2}} \tag{39}$$

Therefore, the steady-state EMSE H_p satisfies

$$\mathrm{H}_p=\frac{\eta_p}{2}Tr(R_{\overline{WX}})\frac{(\mathrm{H}_p+\varsigma_v^2)(\sigma^2+\varsigma_v^2+\mathrm{H}_p)^{3/2}}{(\sigma^2+2\varsigma_v^2+2\mathrm{H}_p)^{3/2}} \tag{40}$$

Theorem 1. In a Gaussian noise environment and with the same step-size, the proposed nonlinear Hammerstein adaptive filter under MCC criterion has a smaller steady-state EMSE than under MSE criterion. As the kernel width increases, their values of the steady-state EMSE will become almost identical.

Proof. It can be shown that [34]

$$\mathrm{H}_{p\text{-}MSE}=\frac{\eta_p Tr(R_{\overline{WX}})\varsigma_v^2}{2-\eta_p Tr(R_{\overline{WX}})} \tag{41}$$

where H_{p-MSE} denotes the steady-state EMSE under MSE criterion. From Equation (40), we have

$$H_p = \frac{\varepsilon \eta_p Tr(R_{\overline{WX}}) \varsigma_v^2}{2 - \varepsilon \eta_p Tr(R_{\overline{WX}})} \tag{42}$$

where $\varepsilon = \frac{(\sigma^2 + \varsigma_v^2 + H_p)^{3/2}}{(\sigma^2 + 2\varsigma_v^2 + 2H_p)^{3/2}}$. Since $\varepsilon < 1$, it holds

$$H_p < H_{p-MSE} \tag{43}$$

Further, as $\sigma \to \infty$, we have $H_p \to H_{p-MSE}$.

Case B. Non-Gaussian Noise

Taking the Taylor series expansion of $f(e(n))$ around $v(n)$ yields

$$\begin{aligned} f(e(n)) &= f(e_p(n) + v(n)) \\ &= f(v(n)) + f'(v(n)) e_p(n) + \frac{1}{2} f''(v(n)) e_p^2(n) + h.o.t \end{aligned} \tag{44}$$

with

$$\begin{aligned} f'(v(n)) &= \exp\left(\frac{-v^2(n)}{2\sigma^2}\right)\left(1 - \frac{v^2(n)}{\sigma^2}\right) \\ f''(v(n)) &= \exp\left(\frac{-v^2(n)}{2\sigma^2}\right)\left(\frac{v^3(n)}{\sigma^4} - \frac{3v(n)}{\sigma^2}\right) \end{aligned} \tag{45}$$

Under the assumptions (A) and (B), we get [34]

$$\lim_{n\to\infty} E\left[f(e(n)) \cdot e_p(n)\right] \approx E\left[f'(v(n))\right] H_p \tag{46}$$

$$\lim_{n\to\infty} E\left[f^2(e(n))\right] \approx E\left[f^2(v(n))\right] + E\left[f(v(n))f''(v(n)) + \left|f'(v(n))\right|^2\right]\mathrm{H}_p \tag{47}$$

Substituting Equations (46) and (47) into Equation (36), we have

$$\mathrm{H}_p = \frac{\eta_p Tr(R_{\overline{\pi}X})E\left[f^2(v(n))\right]}{2E\left[f'(v(n))\right] - \eta_p Tr(R_{\overline{\pi}X})E\left[f(v(n))f''(v(n)) + \left|f'(v(n))\right|^2\right]} \tag{48}$$

Further, substituting Equation (45) into Equation (48), we obtain

$$\mathrm{H}_p = \frac{\eta_p Tr(R_{\overline{\pi}X})E\left[\exp\left(\frac{-v^2(n)}{\sigma^2}\right)v^2(n)\right]}{2E\left[\exp\left(\frac{-v^2(n)}{2\sigma^2}\right)\left(1-\frac{v^2(n)}{\sigma^2}\right)\right] - \eta_p Tr(R_{\overline{\pi}X})E\left[\exp\left(\frac{-v^2(n)}{\sigma^2}\right)\left(1+\frac{2v^4(n)}{\sigma^4}-\frac{5v^2(n)}{\sigma^2}\right)\right]} \tag{49}$$

When only the linear filter with parameter vector $\mathbf{w}(n)$ is adapted, we get

$$2\cdot\lim_{n\to\infty} E\left[f(e(n))\cdot e_w(n)\right] = \eta_p Tr(R_{XP})\lim_{n\to\infty} E\left[f^2(e(n))\right] \tag{50}$$

where $e_w(n) \approx \tilde{\mathrm{w}}^T(n)\mathrm{X}^T(n)\mathrm{p}(n)$, $\tilde{\mathrm{w}}(n) = \mathrm{w}^* - \mathrm{w}(n)$. For Gaussian noise case, we obtain

$$\mathrm{H}_w = \frac{\eta_w}{2}Tr(R_{XP})\frac{(\mathrm{H}_w+\varsigma_v^2)(\sigma^2+\varsigma_v^2+\mathrm{H}_w)^{3/2}}{(\sigma^2+2\varsigma_v^2+2\mathrm{H}_w)^{3/2}} \tag{51}$$

In non-Gaussian environments, we have

$$\mathrm{H}_w = \frac{\eta_w Tr(R_{XP})E\left[\exp\left(\frac{-v^2(n)}{\sigma^2}\right)v^2(n)\right]}{2E\left[\exp\left(\frac{-v^2(n)}{2\sigma^2}\right)\left(1-\frac{v^2(n)}{\sigma^2}\right)\right] - \eta_w Tr(R_{XP})E\left[\exp\left(\frac{-v^2(n)}{\sigma^2}\right)\left(1+\frac{2v^4(n)}{\sigma^4}-\frac{5v^2(n)}{\sigma^2}\right)\right]} \tag{52}$$

Theorem 2. The H_{pw} satisfies the following condition

$$H_{pw} \geq H_p + H_w \tag{53}$$

Proof. Using Equation (31), we derive

$$\begin{aligned} e_{pw}(n) &= w^{*T} X^T(n) p^* - w^T(n) X^T(n) p(n) \\ &= \left(w^{*T} X^T(n) p^* - w^{*T}(n) X^T(n) p(n) \right) + \left(w^{*T} X^T(n) p(n) - w^T(n) X^T(n) p(n) \right) \\ &\approx w^T(n) X^T(n) \tilde{p}(n) + \tilde{w}^T(n) X^T(n) p(n) \\ &= e_p(n) + e_w(n) \end{aligned} \tag{54}$$

It follows that

$$\begin{aligned} H_{pw} &= \lim_{n\to\infty} E\left[e^2{}_{pw}(n) \right] \approx \lim_{n\to\infty} E\left[\left(e_p(n) + e_w(n) \right)^2 \right] \\ &= \lim_{n\to\infty} E\left[e_p^2(n) \right] + 2 \lim_{n\to\infty} E\left[e_p(n) e_w(n) \right] + \lim_{n\to\infty} E\left[e_w^2(n) \right] \\ &= H_p + H_w + 2H_{cross} \end{aligned} \tag{55}$$

where $H_{cross} = \lim_{n\to\infty} E\left[e_p(n) e_w(n) \right]$ stands for the *cross-EMSE* and $H_{cross} \geq 0$ ($H_{cross} = 0$ when $e_p(n)$ and $e_w(n)$ are statistically independent and zero mean) [7]. Therefore, $H_{pw} \geq H_p + H_w$, which completes the proof.

4. SIMULATION RESULTS

Now, we present simulation results to demonstrate the performance of the Hammerstein adaptive filtering under MCC. In order to show the performance of the proposed algorithm in non-Gaussian noises, we adopt the alpha-stable distribution to generate the disturbance noise, whose characteristic function is [32,46]

$$f(\mathrm{t}) = \exp\{\mathrm{j}\delta\mathrm{t} - \gamma|t|^{\alpha}[1 + \mathrm{j}\beta\,\mathrm{sgn}(\mathrm{t})\mathrm{S}(\mathrm{t},\alpha)]\} \tag{56}$$

in which

$$\mathrm{S}(\mathrm{t},\alpha) = \begin{cases} \tan\dfrac{\alpha\pi}{2} & if\,\alpha \neq 1 \\ \dfrac{2}{\pi}\log|t| & if\,\alpha = 1 \end{cases} \tag{57}$$

where $\alpha \in (0, 2]$ denotes the characteristic factor, $-\infty < \delta < +\infty$ is the location parameter, $\beta \in [-1, 1]$ stands for the symmetry parameter, and $\gamma > 0$ is the dispersion parameter. The characteristic factor α measures the tail heaviness of the distribution. The smaller α is, the heavier the tail is. In addition, γ measures the dispersion of the distribution. The distribution is symmetric about its location δ when $\beta = 0$. Such a distribution is called a *symmetric alpha-stable* (*S*α*S*) distribution. The parameters vector of the noise model is defined as $V = (\alpha, \beta, \gamma, \delta)$.

In the simulations below, the input signal considered is a colored signal obtained from the following equation:

$$x(n) = ax(n-1) + \sqrt{1-a^2}\xi(n) \tag{58}$$

with a = 0.95, and $\xi(n)$ being a white Gaussian signal of unit variance. In addition, the coefficient vectors are initialized with the first coefficient equal to 1 and the others equal to zero [7].

4.1. Experiment 1

First, we consider an unknown Hammerstein system with parameter vectors $\mathbf{p}^* = [1, 0.6]$, $\mathbf{w}^* = [1, 0.6, 0.1, -0.2, -0.06, 0.04, 0.02, -0.03, -0.02, 0.01]$. Thus, $M = 2$, $N = 10$. The kernel width σ is 1.0. The noise vector V is set at (1.2, 0, 0.6, 0), and the noise signal is shown in

Figure 3. Simulation results are averaged over 100 independent Monte Carlo runs, and in each simulation, 15,000 iterations are run to ensure the algorithm will reach the steady state, and the steady-state MSE is obtained as an average over the last 2000 iterations. The step-sizes are set at $\mu_p = \mu_w = 0.005$ and $\mu_p = 0.01$, $\mu_w = 0.01$ for MSE and MCC, respectively. Figure 4 shows the average convergence curves under MCC and MSE. As we can see, the Hammerstein adaptive filtering under MCC criterion achieves faster convergence speed and lower steady-state testing MSE than under MSE criterion. Here the testing MSE is evaluated on a test set with 100 samples.

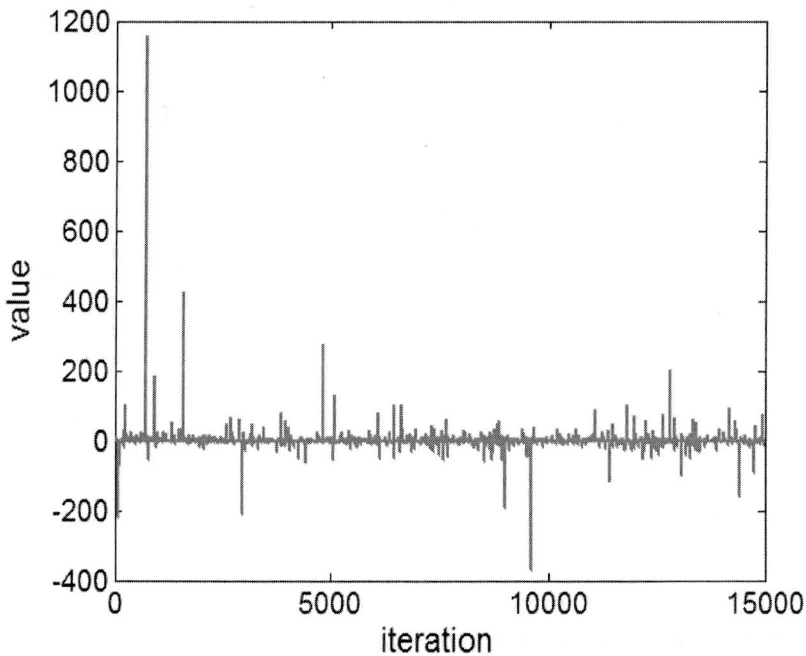

Figure 3. A typical sequence of the alpha-stable noise with $V = (1.2, 0, 0.6, 0)$.

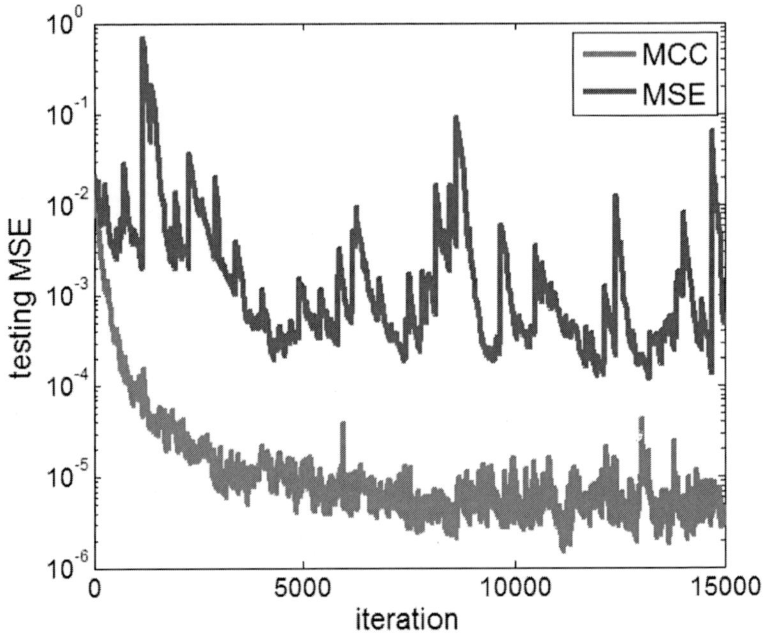

Figure 4. Convergence curves under maximum correntropy criterion (MCC) and mean square error (MSE) (for unknown system with polynomial nonlinearity).

Second, we investigate the performance of the algorithms with different noise parameters. The steady-state MSEs with different γ (0.2, 0.4, 0.6, 0.8, 1.0, 1.2, 1.4, 1.6) and different α (0.2, 0.4, 0.6, 0.8, 1, 1.2, 1.4, 1.6, 1.8, 2.0) are shown in Figure 5and Figure 6, respectively. We observe: (1) In most cases, the new algorithm performs better and achieves a lower steady-state MSE compared with the Hammerstein adaptive filtering under MSE criterion; (2) When α is close to 2.0, the Hammerstein adaptive filtering under MSE criterion can achieve better performance than under MCC criterion. The main reason for this is that, when $\alpha \approx 2.0$, the noise will be approximately Gaussian. Simulation results suggest that the proposed algorithm is particularly useful for identifying a Hammerstein system in non-Gaussian noises.

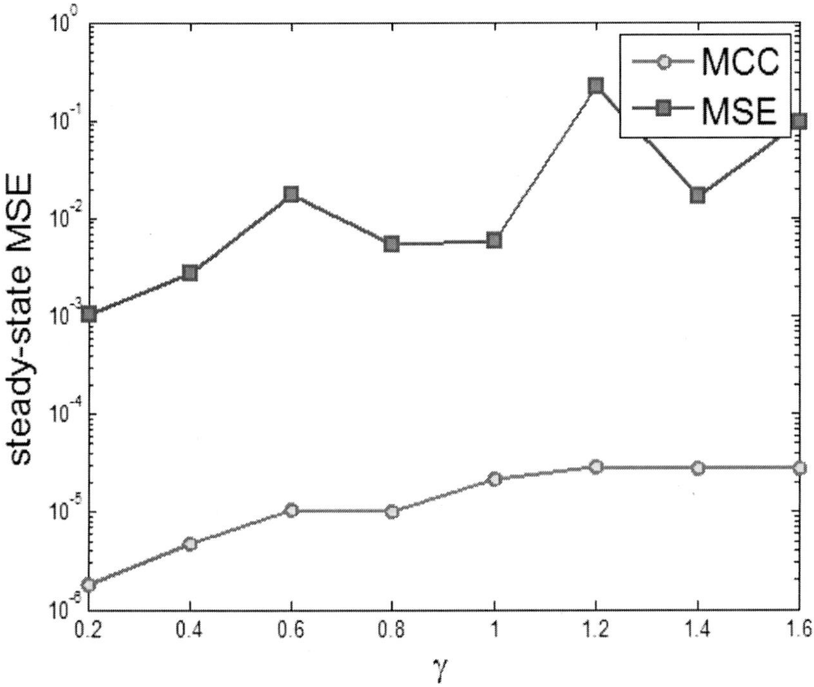

Figure 5. Steady-state mean square error (MSE) with different γ (α = 1.2).

4.2. Experiment 2

The second experiment is drawn from [47]. The nonlinear dynamic system is composed of two blocks. The first block is a non-polynomial nonlinearity

$$s(n) = \sqrt[3]{x(n)} \tag{59}$$

while the second block is an FIR filter with weight vector

$$h = [1\ 0.75\ 0.5\ 0.25\ 0\ -0.25]^T \tag{60}$$

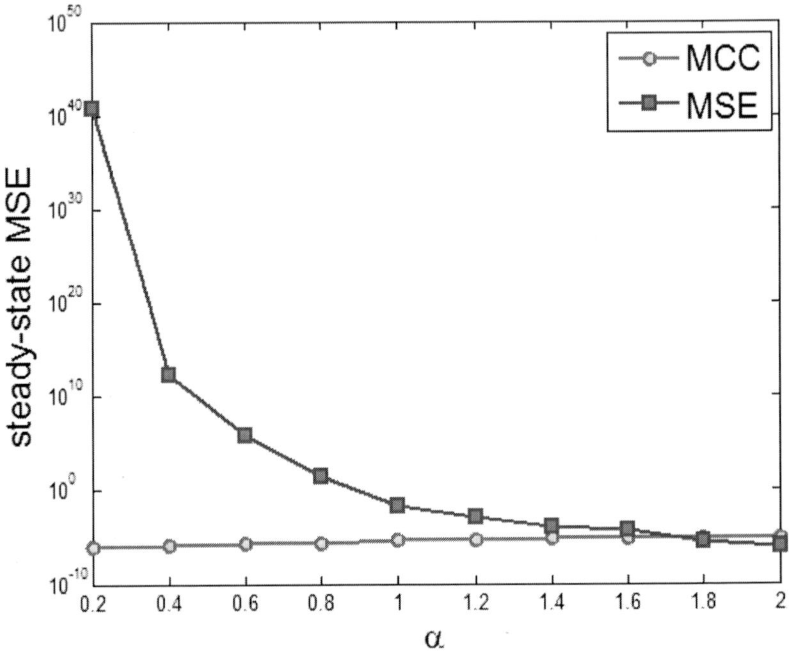

Figure 6. Steady-state mean square error (MSE) with different α (γ = 0.6).

The noise vector V is set at (1.0, 0, 0.8, 0) (see Figure 7 for a typical sequence of the noise), and the polynomial order M and the FIR memory size N are set at 3 and 6, respectively. Simulation results are averaged over 50 independent Monte Carlo runs, and in each simulation, 30,000 iterations are run to ensure the algorithm will reach the steady state, and the steady-state MSE is obtained as an average over the last 2000 iterations. The testing MSE is evaluated on a test set with 100 samples.Figure 8 demonstrates the convergence curves under MCC and MSE. For both adaptive filtering algorithms, the step-sizes are set at $\mu_p = 0.005$, $\mu_w = 0.015$. It can be seen that, the Hammerstein adaptive filter under MCC criterion performs better (say, with faster convergence speed and smaller mismatch error) than under MSE criterion.

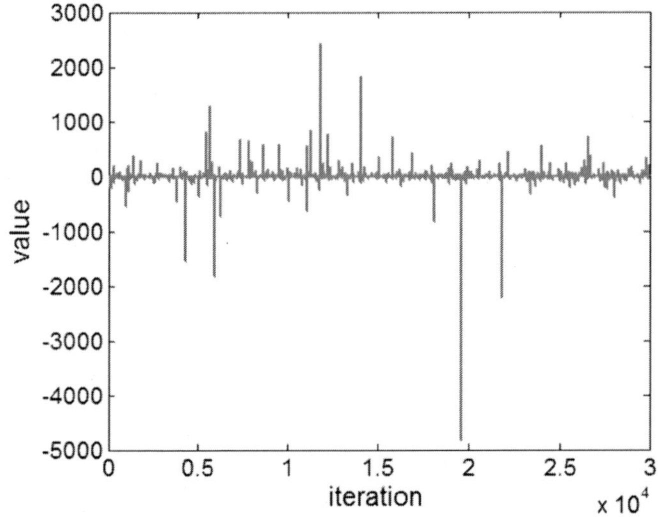

Figure 7. A typical sequence of the alpha-stable noise with V = (1.0, 0, 0.8, 0).

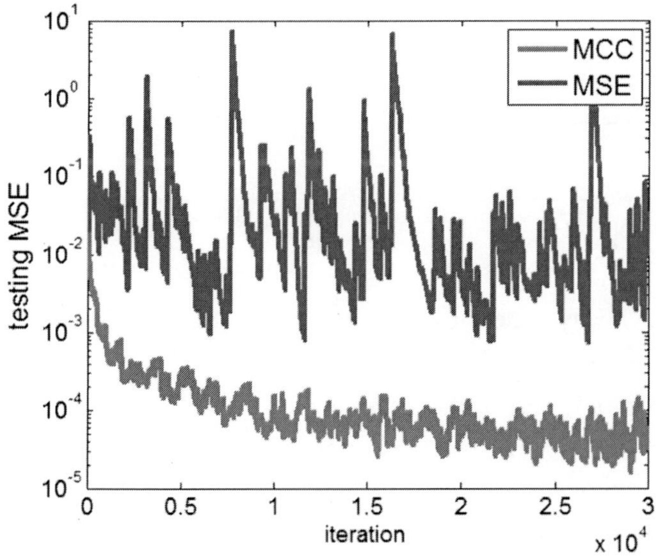

Figure 8. Convergence curves under maximum correntropy criterion (MCC) and mean square error (MSE) (for unknown system with non-polynomial nonlinearity).

Finally, we show the steady-state performance of the algorithms with different kernel widths σ (0.01, 1.0, 2.0, 3.0, 4.0, 5.0). Simulation results are shown in Figure 9. As we can see, the kernel width has significant influence on the performance of the proposed algorithm. In this example, the lowest steady-state MSE is obtained when σ = 1.0.

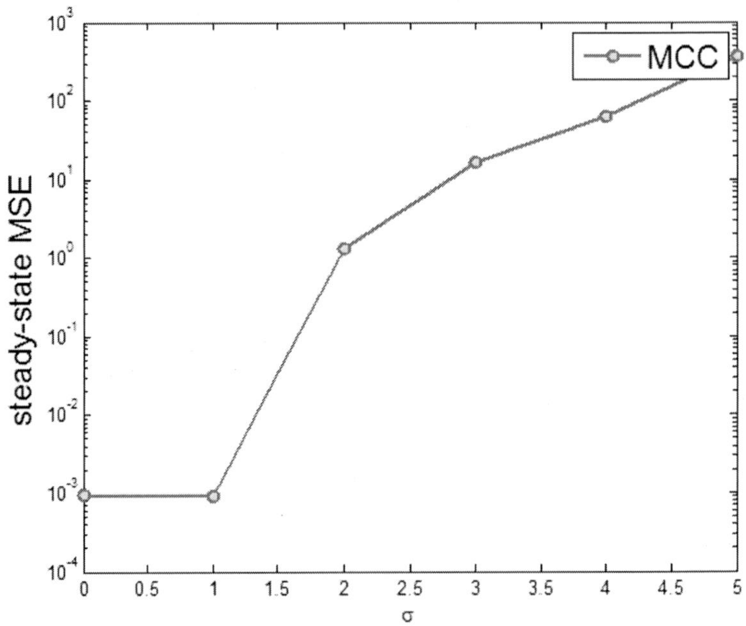

Figure 9. Steady-state mean square error (MSE) with different kernel widths.

5. CONCLUSIONS

The MCC has been successfully applied in domains of machine learning and signal processing due to its strong robustness in impulsive non-Gaussian situations. In this work, we develop a robust Hammerstein adaptive filter under MCC criterion. Different from the traditional Hammerstein adaptive filtering algorithms, the new algorithm use the MCC instead of the well-known MSE as the adaptation criterion, which can achieve desirable performance especially in impulsive noises. Based on [7,31], we carry out the convergence analysis, and obtain some important theoretical results. Simulation examples confirm the excellent performance of the proposed algorithm. How to verify the derived theoretical results is an interesting topic for future study.

ACKNOWLEDGMENTS

This work was supported by 973 Program (No. 2015CB351703) and National Natural Science Foundation of China (No. 61372152, No. 61271210).

AUTHOR CONTRIBUTIONS

Zongze Wu derived the algorithm and wrote the draft; Siyuan Peng proved the convergence properties and performed the simulations; Badong Chen proposed the main idea and polished the language; Haiquan Zhao was in charge of technical checking. All authors have read and approved the final manuscript.

CONFLICTS OF INTEREST

The authors declare no conflict of interest.

REFERENCES

1. Ljung, L. *System Identification—Theory for the User*, 2nd ed.; Prentice Hall: Upper Saddle River, NJ, USA, 1999.

2. Sayed, A.H. *Fundamentals of Adaptive Filtering*; Wiley: Hoboken, NJ, USA, 2003.

3. Wiener, N. *Nonlinear Problems in Random Theory*; MIT Press: Cambridge, MA, USA, 1958.

4. Scarpiniti, M.; Comminiello, D.; Parisi, R.; Uncini, A. Nonlinear spline adaptive filtering. *Signal Process.* **2013**, *93*, 772–783.

5. Bai, E.W. Frequency domain identification of Wiener models. *Automatica* **2003**, *39*, 1521–1530.

6. Ogunfunmi, T. *Adaptive Nonlinear System Identification: The Volterra and Wiener Model Approaches*; Springer: Berlin/ Heidelberg, Germany, 2007.

7. Scarpiniti, M.; Comminiello, D.; Parisi, R.; Uncini, A. Hammerstein uniform cubic spline adaptive filters: Learning and convergence properties. *Signal Process.* **2014**, *100*, 112–123.

8. Ortiz Batista, E.L.; Seara, R. A new perspective on the convergence and stability of NLMS Hammerstein filters. In Proceedings of the IEEE 8th International Symposium on Image and Signal Processing and Analysis (ISPA), Trieste, Italy, 4–6 September 2013; pp. 343–348.

9. Bai, E.W.; Li, D. Convergence of the iterative Hammerstein system identification algorithm. *IEEE Trans. Autom. Control* **2004**, *49*, 1929–1940.

10. Umoh, I.; Ogunfunmi, T. An Affine-Projection-Based Algorithm for Identification of Nonlinear Hammerstein Systems.*Signal Process.* **2010**, *90*, 2020–2030.

11. Umoh, I.; Ogunfunmi, T. An Adaptive Nonlinear Filter for System Identification. *EURASIP J. Adv. Signal Process.* **2009**,*2009*.

12. Umoh, I.; Ogunfunmi, T. An adaptive algorithm for Hammerstein filter system identification. In Proceedings of the EURASIP European Signal Processing Conference, Lausanne, Switzerland, 25–29 August 2008; pp. 1–5.

13. Greblicki, W. Continuous time Hammerstein system identification. *IEEE Trans. Autom. Control* **2000**, *45*, 1232–1236.

14. Jeraj, J.; Mathews, V.J. Stochastic mean-square performance analysis of an adaptive Hammerstein filter. *IEEE Trans. Signal Process.* **2006**, *54*, 2168–2177.

15. Voros, J. Iterative algorithm for parameter identification of Hammerstein systems with two-segment nonlinearities.*IEEE Trans. Autom. Control* **1999**, *44*, 2145–2149.

16. Greblicki, W. Stochastic approximation in nonparametric identification of Hammerstein systems. *IEEE Trans. Autom. Control* **2002**, *47*, 1800–1810.

17. Ding, F.; Liu, X.P.; Liu, G. Identification methods for Hammerstein nonlinear systems. *Dig. Signal Process.* **2011**, *21*, 215–238.

18. Jeraj, J.; Matthews, V.J. A stable adaptive Hammerstein filter employing partial orthogonalization of the input signals.*IEEE Trans. Signal Process.* **2006**, *54*, 1412–1420.

19. Nordsjo, A.E.; Zetterberg, L.H. Identification of certain time-varying nonlinear Wiener and Hammerstein systems.*IEEE Trans. Signal Process.* **2001**, *49*, 577–592.

20. Atiya, A.; Parlos, A. Nonlinear system identification using spatiotemporal neural networks. In Proceedings of the

International Joint Conference on Neural Networks, Baltimore, MD, USA, 7–11 June 1992; pp. 504–509.

21. Volterra, V. *Theory of Functionals and of Integral and Integro-Differential Equations*; Courier Corporation: North Chelmsford, MA, USA, 2005.

22. Principe, J.C.; Liu, W.; Haykin, S. *Kernel Adaptive Filtering: A Comprehensive Introduction*; Wiley: Hoboken, NJ, USA, 2011.

23. Chen, B.; Zhao, S.; Zhu, P.; Principe, J.C. Quantized kernel least mean square algorithm. *IEEE Trans. Neural Netw. Learn. Syst.* **2012**, *23*, 22–32.

24. Chen, B.; Zhao, S.; Zhu, P.; Principe, J.C. Quantized kernel recursive least squares algorithm. *IEEE Trans. Neural Netw. Learn. Syst.* **2013**, *24*, 1484–1491.

25. Chen, B.; Zhao, S.; Zhu, P.; Principe, J.C. Mean square convergence analysis of the kernel least mean square algorithm.*Signal Process.* **2012**, *92*, 2624–2632.

26. Bai, E.W. A blind approach to Hammerstein model identification. *IEEE Trans. Signal Process.* **2002**, *50*, 1610–1619.

27. Stenger, A.; Kellermann, W. Adaptation of a memoryless preprocessor for nonlinear acoustic echo cancelling. *Signal Process.* **2000**, *80*, 1747–1760.

28. Shi, K.; Ma, X.; Zhou, G.T. An efficient acoustic echo cancellation design for systems with long room impulses and nonlinear loudspeakers. *Signal Process.* **2009**, *89*, 121–132.

29. Scarpiniti, M.; Comminiello, D.; Parisi, R.; Uncini, A. Comparison of Hammerstein and Wiener systems for nonlinear acoustic echo cancelers in reverberant environments. In Proceedings of the 17th International Conference on Digital Signal Processing (DSP), Corfu, Greece, 6–8 July 2011; pp. 1–6.

30. Kailath, T.; Sayed, A.H.; Hassibi, B. *Linear Estimation*; Prentice Hall: Upper Saddle River, NJ, USA, 2000.

31. Plataniotis, K.N.; Androutsos, D.; Venetsanopoulos, A.N. Nonlinear filtering of non-Gaussian noise. *J. Intell. Robot. Syst.* **1997**, *19*, 207–231.

32. Weng, B.; Barner, K.E. Nonlinear system identification in impulsive environments. *IEEE Trans. Signal Process.* **2005**, *53*, 2588–2594.

33. Chen, B.; Zhu, Y.; Hu, J.; Principe, J.C. *System Parameter Identification: Information Criteria and Algorithms*; Elsevier: Amsterdam, The Netherlands, 2013.

34. Chen, B.; Xing, L.; Liang, J.; Zheng, N.; Principe, J.C. Steady-state Mean-square Error Analysis for Adaptive Filtering under the Maximum Correntropy Criterion. *IEEE Signal Process. Lett.* **2014**, *21*, 880–884.

35. Principe, J.C. *Information Theoretic Learning: Renyi's Entropy and Kernel Perspectives*; Springer: New York, NY, USA, 2010.

36. Singh, A.; Principe, J.C. Using Correntropy as a cost function in linear adaptive filters. In Proceedings of the IEEE International Joint Conference on Neural Networks (IJCNN), Atlanta, GA, USA, 14–19 June 2009; pp. 2950–2955.

37. Chen, B.; Principe, J.C. Maximum correntropy estimation is a smoothed MAP estimation. *IEEE Signal Process. Lett.***2012**, *19*, 491–494.

38. Zhao, S.; Chen, B.; Principe, J.C. Kernel adaptive filtering with maximum correntropy criterion. In Proceedings of the IEEE International Joint Conference on Neural Networks (IJCNN), Killarney, Ireland, 31 July–5 August 2011; pp. 2012–2017.

39. Ogunfunmi, T.; Paul, T. The Quaternion Maximum Correntropy Algorithm. *IEEE Trans. Circuits Syst. -II (TCAS-II)***2015**, *62*, 598–602.

40. Mandic, D.P.; Hanna, A.I.; Razaz, M. A normalized gradient descent algorithm for nonlinear adaptive filters using a gradient adaptive step size. *IEEE Signal Process. Lett.* **2001**, *8*, 295–297.

41. Al-Naffouri, T.Y.; Sayed, A.H. Adaptive filters with error non-linearities: Mean-square analysis and optimum design.*EURASIP J. Appl. Signal Process.* **2001**, *4*, 192–205.

42. Lin, B.; He, R.; Wang, X.; Wang, B. The steady-state mean-square error analysis for least mean p-order algorithm. *IEEE Signal Process. Lett.* **2009**, *16*, 176–179.

43. Yousef, N.R.; Sayed, A.H. A unified approach to the steady-state and tracking analysis of adaptive filters. *IEEE Trans. Signal Process.* **2001**, *49*, 314–324.

44. Duttweiler, D.L. Adaptive filter performance with nonlinearities in the correlation multiplier. *IEEE Trans. Acoust. Speech Signal Process.* **1982**, *30*, 578–586.

45. Mathews, V.J.; Cho, S.H. Improved convergence analysis of stochastic gradient adaptive filters using the sign algorithm. *IEEE Trans. Acoust. Speech Signal Process.* **1987**, *35*, 450–454.

46. Shao, M.; Nikias, C.L. Signal processing with fractional lower order moments: Stable processes and their applications.*Proc. IEEE* **1993**, *81*, 986–1010.

47. Hasiewicz, Z.; Pawlak, M.; Sliwinski, P. Nonparametric identification of nonlinearities in block-oriented systems by orthogonal wavelets with compact support. *IEEE Trans. Circuits Syst. I Regul.* 2005, *52*, 427–442.

Index

A

C

D

E

F

G

H

K

L

M

N

O

P

R

S

T

U

W